SpringerBriefs in Earth System Sciences

SpringerBriefs in Earth System Sciences present concise summaries of cutting-edge research and practical applications. The series focuses on interdisciplinary research linking the lithosphere, atmosphere, biosphere, cryosphere, and hydrosphere building the system earth. It publishes peer-reviewed monographs under the editorial supervision of an international advisory board with the aim to publish 8 to 12 weeks after acceptance. Featuring compact volumes of 50 to 125 pages (approx. 20,000—70,000 words), the series covers a range of content from professional to academic such as:

- A timely reports of state-of-the art analytical techniques
- bridges between new research results
- snapshots of hot and/or emerging topics
- literature reviews
- in-depth case studies

Briefs are published as part of Springer's eBook collection, with millions of users worldwide. In addition, Briefs are available for individual print and electronic purchase. Briefs are characterized by fast, global electronic dissemination, standard publishing contracts, easy-to-use manuscript preparation and formatting guidelines, and expedited production schedules.

Both solicited and unsolicited manuscripts are considered for publication in this series.

M. Daud Rafiqpoor · Siegmar-W. Breckle

The World's Ecoclimates

A New Digital Climate Classification and Its Importance for Modelling Future Climate Change

Springer

M. Daud Rafiqpoor
Department I: Biodiversity of Plants
Bonn Institute for Organismic Biology
University Bonn
Bonn, Germany

Siegmar-W. Breckle
Department of Ecology
Bielefeld, Nordrhein-Westfalen, Germany

ISSN 2191-589X ISSN 2191-5903 (electronic)
SpringerBriefs in Earth System Sciences
ISBN 978-3-032-06455-4 ISBN 978-3-032-06456-1 (eBook)
https://doi.org/10.1007/978-3-032-06456-1

This Springer imprint is published by the registered company Springer Nature Switzerland AG
The registered company address is: Gewerbestrasse 11, 6330 Cham, Switzerland

Foreword

Global climate change is one of humanity's most critical problems. Researchers often refer to climate change, biodiversity loss and environmental pollution as interconnected environmental emergencies threatening well-being of our current and future generations. These crises should be addressed together. With the state of the environment continuously worsening, scientific literacy and capacity building have become key concerns for evidence-based decision-making and an informed public. In short, we need to *Make Peace with Nature*, as a report of the UN's Environment Programme (UNEP) is titled.

The World's Ecoclimates by M. Daud Rafiqpoor and Siegmar-W. Breckle takes us on a journey back to the roots of the climate debate: How has climate been perceived since about 1900? How do we classify *eco*climate? How can we do this in modern digital format (Part I) and, in consequence, how can this information contribute to improving modelling approaches for future climate change scenarios (Part II)?

The book suggests a transformation from traditional approaches to an *eco*climatic view of how to classify climate. The authors draw up a dynamic model, from classifications developed over a century ago to a more holistic scheme centered on potential evapotranspiration as key process shaping Earth's vegetation patterns. Most importantly, the new *Digital Ecoclimate Classification System*, based on high-resolution global data sets, allows for estimating global shifts in climatic zones and concomitant changes in vegetation patterns due to climate change. The authors' approach may not only be developed into a powerful tool for climate science and its applications, but also be useful for interferring changes in the geography of biological diversity.

Rafiqpoor and Breckle have collaborated for many years, well-illustrated in their impressive list of scientific publications, which includes a substantial number of co-authored works. One of their geographical priorities is South Asia and in particular Afghanistan. Most notable are their *Bibliography of Botanical and Ecological Literature for Afghanistan and Adjacent Countries* and their landmark book publication *Vegetation and Climate*. Both titles have also been published in Dari, within the framework of a project on textbooks for universities in Afghanistan-A role model initiative of environmental education for the Global South, supported by the *German Academic Exchange Service* (DAAD) and the *Gesellschaft für Internationale Zusammenarbeit* (GIZ).

Indeed, *The World's Ecoclimates* may contribute to building the capacities so desperately needed, in particular in fields related to climate, vegetation, biodiversity, and global climate change. Students should learn about biogeography, i.e. about the patterns and processes which have shaped the distribution of life on Earth, including the human species and its planet-wide impacts on the environment. Of particular importance in this context are latest insights from Earth system science, holistic thinking, and the role of entities of the United Nations system. As regards efforts to tackle man-made climate change, most relevant are the UN Framework Convention on Climate Change (UNFCCC) and the Intergovernmental Panel on Climate Change (IPCC). As discussed in Part II of the book, ecoclimatic classifications may be instrumental for IPCC's emission scenarios (SRES) for the middle and the end of the 21st century, respectively.

As stated in UNESCO's *World in 2030* public survey report: "*UNESCO aims to enable Member States to take urgent action on climate change and its impacts across the Organization's mandate. This involves supporting Member States to develop and implement education and public awareness programmes and policies,*" [...] "*it involves promoting interdisciplinary climate knowledge and scientific cooperation, including in relation to the interaction between biodiversity and climate,*" and "*deals with the interactions between climate, biodiversity, water, oceans, and disaster risk reduction*" (UNESCO 2021).

Respondents to *World in 2030* were to choose up to four global challenges "that they were most worried about for a peaceful society in 2030." Of these, *climate change and loss of biodiversity* was the "top global challenge across all regions and demographics, selected by over two thirds (67%) of global respondents." As regards top solutions to these challenges, *education* "emerged overwhelmingly" as "the number one solution," followed by *science and science-based solutions*, and *international cooperation.*

Isn't this what *The World's Ecoclimates* and its "environment" are all about?

Prof. Dr. Walter R. Erdelen
Former Assistant Director-General
for Natural Sciences
United Nations Educational, Scientific
and Cultural Organization (UNESCO)
Paris, France

Reference

UNESCO (2021) The World in 2030. Public Survey Report. UNESCO, Paris, France

Preface

Neither climate nor weather know political boundaries.

Science is the key to understanding, mitigating and adapting to a globally changing climate.

Climate encompasses a bundle of ecological factors that are very formative for the earth's ecosystems and thus for human life. Climate classification has been the subject of intensive scientific research for decades. Climate zones, vegetation zones, ecotypical classifications and many other approaches attempt to provide a plausible overview on climates of the Earth. Climate is variable, it is subject to many influences; climate can change: Climate classification and the development of a new concept of "ecoclimate classification" is a topic of this book.

In 1824, the French mathematician Joseph Fourier came to the conclusion that certain atmospheric gases have an insulating effect and therefore influence the earth's temperature. This phenomenon was later referred to as the **natural greenhouse gas effect**. In 1895, the Swedish physicist and Nobel Prize winner for chemistry Svante Arrhenius was the first to recognize that this natural effect, which makes life on Earth possible in the first place, can be influenced by humans. He calculated a temperature rise of four to six degrees if CO_2 in the atmosphere were to double. In the last two decades, thanks to the efforts of numerous scientists and their cooperation within the framework of the Intergovernmental Panel on Climate Change (IPCC), the last doubt that humans have been the main causers of climate change since the beginning of industrialization has been dispelled.

Thanks to weather records, we now have a solid knowledge of how global temperatures have changed over the last century and a half. Thousands of years ago, the climate left traces in the environment that can still be analysed today. Leonardo da Vinci already suspected that the annual rings of trees reflect the climatic growth conditions. In addition to growth rings in trees, conclusions can be drawn from ice cores, coral reefs and sediments about the climate or environmental conditions of past times. With these proxies, we can extend our knowledge of the climate of the past by hundreds of thousands to millions of years.

Using climate models that simulate the Earth's climate system, researchers can reliably reconstruct past climates and project future developments. The projections are based on various assumptions about the future development of factors that influence the climate.

In the 1960s, climate models were still in their infancy. Since then, however, they have been continuously developed and improved. The quality of climate models is demonstrated, among other things, by the extent to which projections match the reality observed later. However, due to the complexity of the climate system, deviations can never be completely ruled out. Nevertheless, computer models can now reliably simulate global, long-term climate trends.

The book discusses various climate change scenarios and uses the *new digital ecoclimate classification* to illustrate the *shift in climate zones* and demonstrate the patterns of ecoclimate types *caused by climate change* around the globe. Maps and zonobiome climate diagrams are used to explain and illustrate these patterns, providing clear and rapid visualization.

Bonn, Germany — M. Daud Rafiqpoor
Bielefeld, Germany — Siegmar-W. Breckle

Acknowledgements

We are grateful to Prof. Dr. Holger Kreft (Göttingen) and Dr. Laurens Geffert for GIS support, to Prof. Dr. Walter Erdelen (Former Assistant Director-General for Natural Sciences at UNESCO in Paris) and to Prof. Dr. Wilhelm Barthlott (Former Director of the Nees Institute for Biodiversity of Plants, University of Bonn) and Prof. Dr. Holger Kreft for many valuable comments on the manuscript, and to Dr. Markus Wagner (NW German Forest Research Institute, Department of Environmental Control, Göttingen) for providing Solling data.

Contents

Abbreviations

AR3	(= TAR) Third Assessment Report of IPCC
AR4	(= FAR) Fourth Assessment Report of the IPCC
AR6	(= SAR) Sixth Assessment Report of the IPCC
CCPI	Climate Change Performance Index
CFC	**C**hloro**F**luoro**C**arbons
CMIP	**C**oupled **M**odel **I**ntercomparison **P**roject of World Climate Research Pro-gramme
COP	Conference of the Parties
CRU	Climatic Research Unit (University of East Anglia)
CSR	Capacity for Plant Species Richness
DLV	Day Length Variation
DWD	Deutscher Wetterdienst (German Weather Service)
ETP	Evapotranspiration
FAO	Food and Agriculture Organisation of the United Nations
GCM	General Circulation Modell
GHCN	Global Historical Climatology Network
GHG	Green House Gases
GIS	Geographic Information System
GMST	Global Mean Surface Temperature
GPCC	Global Precipitation Climatology Centre (DWD)
IPCC	The Intergovernmental Panel on Climate Change
ITCZ	Inter Tropical Convergence Zone
K	Kelvin (= Δ°C)
L/V	Bowen-Ratio = Latent Heath/Sensible Heath
N	Precipitation
OB	Orobiome
pET	Potential Evapotranspiration
pLV	Potentialle Landschaftsverdunstung (= potentiell Landscape Evapotranspiration)
PNV	Potential Natural Vegetation
SRES	Special Report on Emissions Scenarios

SSPs	Seven Socio-economic Pathways
WTB	Water Balance
ZB	Zonobiome
ZE	Zonoecotone

Part I
Ecoclimate Classification Based on Heat and Water Budgets

In recent years, several attempts have been made, with different motivations, to reproduce Wladimir Köppen's (1900) climate classification in digital form, based on various temperature and precipitation records, in order to make it available for further use. Unfortunately, by maintaining the original criteria of Köppen's classification, these attempts did not achieve any qualitative progress. Compared to the Köppen's original maps (1900, 1923, 1936) and the later version published by Rudolf Geiger (1961), these "*new*" digital classifications inevitably showed only border adjustments due to the use of different digital data sets with different resolutions.

Recently, many people have expressed the wish to attempt a new digital climate classification based on publicly available, comprehensive, high-resolution global data sets, which would be useful for various ecological analyses. Here, an attempt has been made to present a **Digital Ecoclimate Classification System** based on high-resolution global data sets. Taking into account the ecophysiological requirements of Earth's vegetation cover by considering **potential evapotranspiration** (ETP = evaporation of soils and rocks + transpiration of plants), the present ecoclimate classification provides a spatial arrangement of global climates on ecological basis.

In contrast to previous classifications, the **Ecoclimate Classification** does not primarily use the usual "**climatological averages**", i.e. the mean values of temperature and precipitation and not their indices, but calculates climatic parameters relevant for plant life (i.e., **the length of the thermal and hygric growing seasons**) as ecophysiologically meaningful variables, and as a criterion for the delimitation of climate types, on the basis of publicly available monthly temperature and precipitation data (see below). Thus, it gives answer to many environmental questions and also to the global climate change debate. However, the ecoclimate classification model presented here has the potential to infer changes in vegetation cover from changes in climate and vice versa.

The contributions of **Working Group I** to the Third Assessment Report (TAR or AR3) of the **I**ntergovernmental **P**anel on **C**limate **C**hange (IPCC), for example, predict continued atmospheric warming with an increase in mean annual temperatures of about 1.8–3.5 K by 2050 and 2–7 K by 2100 (Cubasch et al. 2001, IPCC 2023; see also Hobbs et al. 2006; Ohlemüller et al. 2006; Ricciardi 2007; Williams &

Jackson 2007). It is in this context that the **Ecoclimate Classification** is a powerful tool for assessing the dimensions of habitat change for flora and fauna or, conversely, for inferring the shift of ecoclimatic boundaries from changes in vegetation cover caused by climate change (see Part II of this contribution).

References

Cubasch U (2001) Climate change 2001: The scientific basis. Third Assessment Report of IPCC, Chapter 9: 525–582. https://www.researchgate.net/publication/290809956

Geiger R (1961) Überarbeitete Neuausgabe von Köppen/Geiger: Klimate der Erde (Wandkarte 1:16 Mio.). Klett-Perthes, Gotha

Hobbs RJ, Arico S, Aronson J, Baron JS, Bridgewater P, et al. (2006) Novel ecosystems: Theoretical and management aspects of the new ecological world order. Glob Ecol Biogeogr 15(1):1–7. https://doi.org/10.1111/j.1466-822X.2006.00212.x

Köppen W (1900) Versuch einer Klassifikation der Klimate, vorzugsweise nach ihren Beziehungen zur Pflanzenwelt. Geographische Zeitschrift, vol 6: 593–611 und 657–679

Köppen W (1923) Die Klimate der Erde – Grundriss der Klimakunde, Berlin, Leipzig

Köppen W (1936) Das geographische System der Klimate. Handbuch der Klimatologie. Gebrüder Bornträger, Berlin

Lauer W, Rafiqpoor MD (2001) Biodiversität—Geodiversität—Ökodiversität. Die Artenvielfalt—der lebendige Reichtum der Erde. In: Meyers Großer Weltatlas. Atlas zum 21. Jahrhundert. Meyers Lexikonverlag, Mannheim, pp 28–29

Ohlemüller R, Gritti ES, Sykes MT, Thomas CD (2006) Towards European climate risk surfaces: The extent and distribution of analogous and non-analogous climates 1931–2100. Glob Ecol Biogeogr 15:395–405

Ricciardi A (2007) Are modern biological invasions an unprecedented form of global change? Conserv Biol 21:329–336

Williams JW, Jackson ST (2007) Novel climates, no-analog communities, and ecological surprises. Front Ecol Environ 5(9):475–482. https://esajournals.onlinelibrary.wiley.com/doi/10.1890/070037

Chapter 1
Introduction—History

In general, the aim of a **classification** is to organise a set of data about something so that it can be understood effectively and easily and to distinguish different categories. Climate is obviously dominant in controlling the broad-scale distribution of life: species, communities, biomes and key ecosystem processes (Lomolino et al. 2005). As biosynthetic pathways and other biological attributes are affected by light, temperature and water availability, these variables directly or indirectly influence the distribution of individuals, species, communities and biomes (Stephenson 1990). Ecosystem properties such as biomass, primary productivity, leaf area or the composition and dominance of plant functional types and thus species assemblages also change mainly along temperature and moisture gradients (Holdridge 1947, 1959, 1967; Box 1981, 1995). In addition, other important abiotic factors, like tropical storms (hurricanes, tayfuns), but even more floods, droughts and fires, are also largely driven by climate (Bond et al. 2005). Ecologically meaningful climate classifications thus provide a conceptual framework for further studies in biogeography, macroecology, vegetation science and climate change.

It is, however, no doubt, that surface properties and soil conditions, and especially **land use**, have feedbacks on the climatic conditions (Lozan et al. 2021).

Understanding the Earth's climate is such a huge amount of information that it needs to be broken down into smaller areas of interest. Since the beginning of the twentieth century, climatologists have therefore developed a number of classification systems to bring order to the vast amount of information about the Earth's climate.

There are three basic types of climate classifications in use in climatology (Ritter 2003). Most classification systems in use today are based on the system introduced by Köppen (1900). This classification system, in which climates are grouped into five main types, generally coincides with global vegetation patterns according to the concept of De Candolle (1874), who classified the Earth's vegetation cover into five main zones, adapted from the "**Climata**" zones of the Greek astronomer Hipparchus (~190–120 B.C.), namely **A**: the plants of the torrid zone; **B**: the plants of the dry zone; **C**: the plants of the temperate zone; **D** and **E**: the plants of the frigid zone (Sanderson

M. D. Rafiqpoor and S.-W. Breckle, *The World's Ecoclimates*,
SpringerBriefs in Earth System Sciences,
https://doi.org/10.1007/978-3-032-06456-1_1

1999). However, Alexander von Humboldt was the first naturalist who, between 1799 and 1803, collected a vast amount of observations on the geography of plants in the New World. He illustrated his idea in a map and a three-dimensional profile of the Andes (Fig. 1.1), in which he distinguished five thermal elevational belts for the tropical mountains and presented them analogously to the horizontal distribution of climate zones (*tierra caliente, tierra templada, tierra fría, tierra helada, tierra nevada*), corresponding to the elevational arrangement of vegetation belts of the tropical Andes (Humboldt and Bonpland 1805, Delametherie 1805).

In Köppen's classification, the main parameters for differentiating between climate types are the annual mean and seasonal extremes of temperature and precipitation, as well as temperature and precipitation indices as the so-called "**Aridity Index**". The intended climate classification is defined as an "**empirical**" or "**quantitative**" system, in contrast to an "**explanatory-descriptive**" or "**genetic**" classification system, which is based on the genesis (i.e., origin) of climatic phenomena, e.g. on information about the climatic elements of solar radiation, air masses, pressure systems, etc. (Strahler 1971, Strahler and Strahler 2005, Lauer and Klaus 1975, Lauer 1991).

A large number of classifications by German geographers (Alissow 1936, 1954; Kupfer 1954; Neef 1954; Flohn 1957; Hendl 1963; Lauer 1991) belong to the genetic classification system, while the others (Köppen 1900, 1923, 1936; von Wissmann 1939; Holdridge 1947, 1959, 1967; Creutzburg 1950 and Troll 1964) belong to the empirical system. The **third type of classification system** used is that which has been created for, or as an outgrowth of, a particular climate-related problem. For example, the classification systems of Thornthwaite (1943, 1948), Lauer (1952), Lauer and Frankenberg (1985, 1988), Lauer et al. (1996) are based on potential evapotranspiration, and/or the length of thermal and hygric growing seasons (Lauer and Rafiqpoor 2002), respectively, and thus define climates according to water, heat and energy availability for plant formations. Siegmund and Frankenberg (1999) have didactically processed the classification approach of Lauer et al. (1996) into a 'modular system' that takes into account the degree of complexity of the climate classification for specific age and school type-related needs.

Loidi et al. (2022) have structured the world's terrestrial biomes climatically, but also based on an index of so called '*positive temperature*' (Tp) and '*positive precipitation*' (Pp), and presented a world map of biomes on this basis. In this classification, too, the boundaries between different climatic zones are blurred, especially the altitudinal belts of tropical and extra-tropical high mountains. Using global data sets, Mucina and Wasserman (2026) have redefined the Whittaker's (1975) geographical interpretation of "Zonal Schimperian World".

Fig. 1.1 Alexander von Humboldt's scheme of the major vegetation formations of the Earth, with vegetation profiles of some of the world's mountains (Humboldt and Bonpland 1805, Delametherie 1805). From the Atlas to Alexander von Humboldt's Cosmos, p. 31. Downloaded on 11 June 2025 from https://es.pinterest.com/pin/25825397841726969/

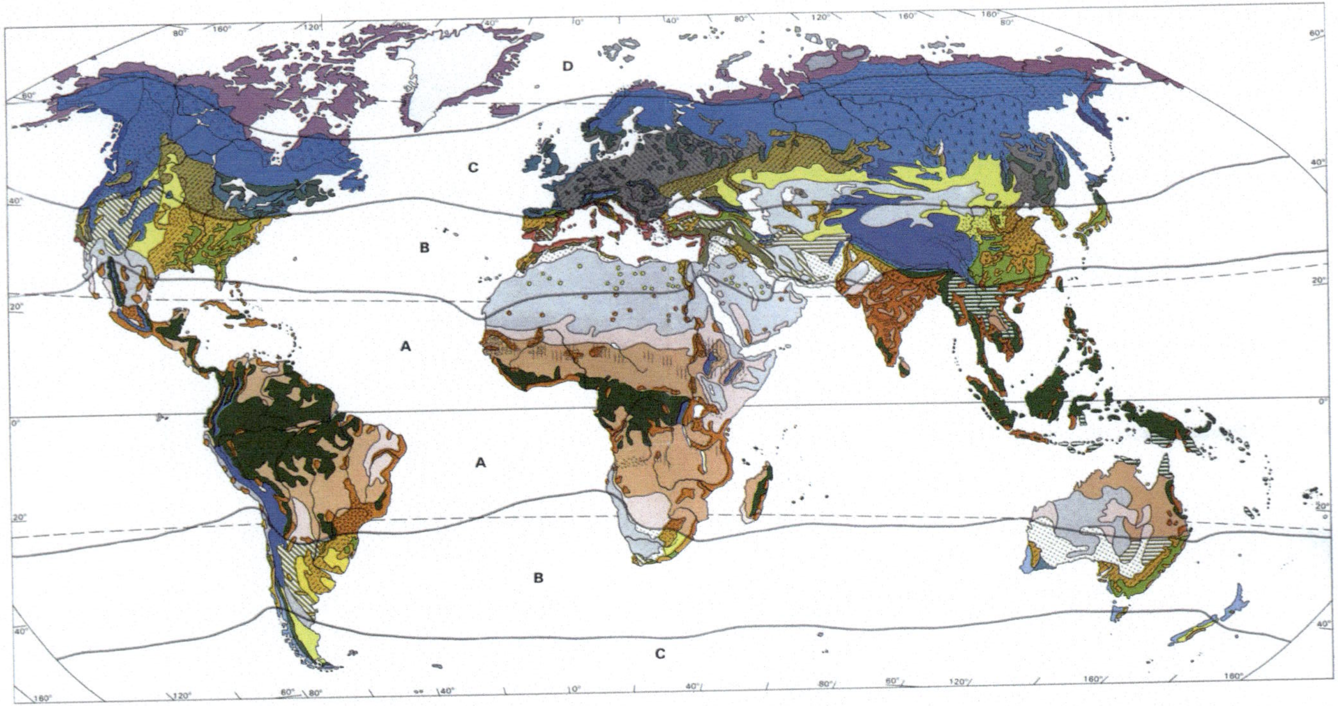

Fig. 1.2 The real vegetation and land use of the Earth. Lines running parallel to the geographical latitudes separate the solar-climatic radiation zones of the Earth A, B, C, D (from Lauer, Rafiqpoor, Frankenberg 1996)

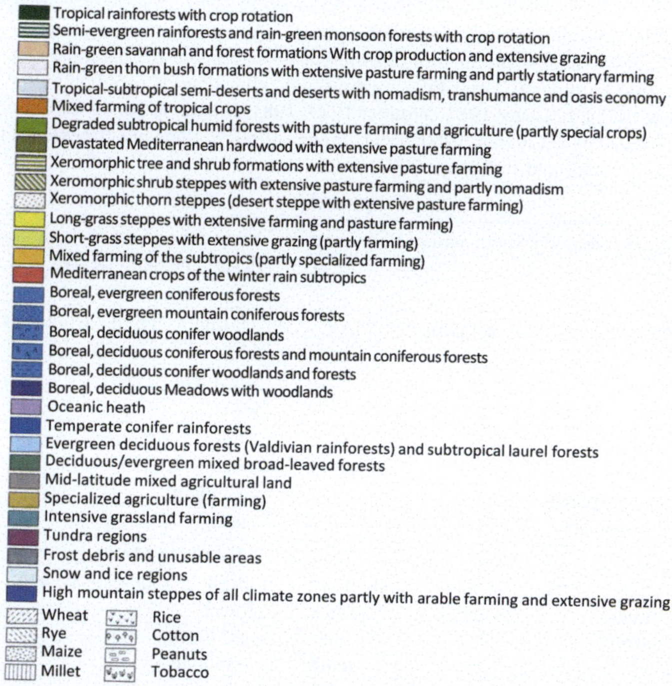

Fig. 1.2 (continued)

Neither the *"new"* digital versions of the Köppen climate classification (i.e., Kottek et al. 2006 or Peel et al. 2007) nor that of Loidi et al. (2022) refer to the Lauer and Rafiqpoor (2002) ecoclimate classification system.

In the ecoclimate classification of Lauer and Rafiqpoor (2002), an attempt was made to calculate the water and heat requirements of vegetation under different environmental conditions. This **ecoclimate classification system** grew out of the problem of trying to predict the heat and water supply and demand of **real vegetation formations** (Fig. 1.2) via thermal and hygric vegetation periods in different climatic regions. Lauer and Rafiqpoor (2002) have used thermal thresholds for natural vegetation formations (Table 1.1) as well as for the characteristic crop types for certain regions of the world (Table 1.2) according to Fig. 1.2.

Table 1.1 The thermal elevational belts of the tropics (A) and the temperature thresholds for the growth and photosynthetic production of different formations of the natural vegetation of the subtropics (B), mid-latitudes (C) and Polar regions (D) (after data of Fröhlich and Willer 1977, Lauer and Klaus 1975a, b, Lauer 1981, Winiger 1979, 1981, Franke 1982, 1985, Rehm and Espig 1984, Geisler 1980, 1983, Larcher 1980, 1994, Walter 1960, Walter and Breckle 1983ff., 1986, 1991, 1994), (from Lauer and Rafiqpoor 2002)

Tropics (A)

Tierra sub-nevada 3 °C	super-páramo	Super-pajonales	Puna brava	Frost debris belt of the high-altitude desert and semi-desert		
Tierra helada 6 °C	Páramo	Moist puna	Dry puna	Thorn and succulent puna	High altitude desert and semi-desert	
Tierra fría 12 °C	Upper montane forest (partly fog-forest)	Moist sierra	Dry sierra	Thorn and succulent sierra	Desert-sierra	High montane desert
Tierra templada 18 °C	Montane rain forest (partly cloud forest)	Montane moist forest and moist savannah	Montane thorn forest and thorn savannah	Montane thorn forest and thorn savannah	Montane desert-savannah	Montane desert
Tierra caliente 27 °C	Lowland rain forest	Moist forest and moist savannah	Dry forest and dry savannah	Thorn forest and thorn savannah	Desert-savannah	Desert

Subtropics (B)		**Mid-Latitudes (C)**	
Vegetation formation	Thermal threshold values (°C)	Vegetation formation	Thermal threshold values (°C)
Subtropical moist forest	12	Boreal coniferous forest	≥5
Coniferous forest	10	Moist coniferous forest	≥5
Summer green deciduous forest	10	Temperate deciduous forest	7
Sclerophyllous forest	12	Mixed deciduous forest	10
Steppes	11	Steppes	10
Pampa	11	Patagonian steppes	7
Semideserts	11	Semideserts	10
Deserts	11	Deserts	10
High mountain formations	6	High mountain formations	5
Cultivated areas	>10	Cultivated areas	>7

(continued)

Table 1.1 (continued)

Polar Regions (D)	
Vegetation formation	Thermal threshold values (°C)
Tundra	≥5
Subpolar frost debris zone	≥3

Table 1.2 Threshold values for the minimum and optimum temperature for growth and photosynthetic production of major crops (after Lauer and Rafiqpoor 2002)

Cultivated plants	**Minimal thermal requirements (°C)**	**Optimal thermal requirements (°C)**
Cocoa-Palm (Cocos nucifera)	24	26–27
Yams (Dioscorea spec.)	20	25–30
Sugar-cane (Saccharum officinarum)	18–20 at 15 °C stops growth	25–28
Maniok (Manihot esculenta)	20	> 27
Cocoa (Theobroma cacao)	> 20	25–28
Coffee (Coffea spec.)	18	> 22
Tea (Camellia sinensis)	18	28
Pineapple (Ananas comosus)	> 18	> 20
Millet (Panicum spec.)	12–15	32–37
Sweet potato (Ipomoea batata)	10	26–30
Tobacco (Nicotiana tabacum)	> 15	25–35
Rice (Oryza sativa)	12–18	30–32
Olive-tree (Olea europaea)	12–15	18–22
Sesam (Sesamum indicum)	12–15	25–27
Gourd (Cucurbita spec.)	> 15	37–40
Cotton (Gossypium spec.)	18	30
Peanut (Arachis hypogaea)	15	30
Maize (Zea mays)	12–15	30–35
Potato (Solanum tuberosum)	8–10	16–24
Winter wheat (Triticum aestivum)	4–6	15–30
Rye (Secale cereale)	4–6	25–30
Barley (Hordeum vulgare)	4–6	20–25
Oats (Avena sativa)	4–6	25–30
Sugar- and Beta-beet (Beta vulgare)	4–5	20–25
Winter grain	4–6	20–30
Summer grain	6–8	20–25

References

Alissow BP (1936) Die geographischen Typen der Klimate. Meteorol Gidrol 6:16–25
Alissow BP (1954) Die Klimate der Erde, Russian original Edition 1950, Berlin
Bond WJ, Woodward FI, Midgley GF (2005) The global distribution of ecosystems in a world without fire. New Phytol 165:525–538
Box EO (1981) Macroclimate and plant forms: an introduction to predictive modelling in phytogeography. Publishers, The Hague, BV
Box EO (1995) Factors determining distributions of tree species and plant functional types. Vegetatio 121:101–116. https://doi.org/10.1007/BF00044676
Creutzburg N (1950) Klima, Klimatypen und Klimakarten. Petermanns Geogr Mitt 49:59–69
De Candolle A (1874) Constitution dans le végétal de groups physiologiques applicable á la géographie botanique ancienne et modern. Univers, Arch. d. Scien. Bibl
Delametherie JC (1805) Reisen der Herren v. Humboldt und Bonpland nach den Wendekreisen in den Jahren 1977, 1800, 1801, 1802, 1803 und 1804. Ein Auszug aus ihren Memoiren. Aus dem Französischen. Bayer und Waring, Erfurt
Flohn H (1957) Zur Frage der Einteilung der Klimazonen. Erdkunde 11:161–175
Franke G (1982) Nutzpflanzen der Tropen und Subtropen. Leipzig
Franke G (1985) Nutzpflanzenkunde. Nutzbare Gewächse der gemäßigten Breiten, Subtropen und Tropen. Stuttgart, New York
Fröhlich W, Willer M (1977) Bodentemperaturen und obere Waldgrenze. Vorläufiger Bericht über Untersuchungen an den Hochbergen Oszafrikas. Die Erde 108:347–355
Geisler G (1980, 1983) Pflantenbau. Ein Lehrbuch - Biologische Grundlagen und Techniken der Pflanzenproduktion. Verlag Paul Parex Berlin/Hamburg. ISBN: 3489604105
Geisler G (1988a) Ertragsbildung von Kulturpflanzen des gemäßigten Klimas. Erträge der Forschung Band 149. Wissenschaftliche Buchgesellschaft Darmstadt.
Geisler G (1988b) Pflanzenbau. Ein Lehrbuch—Biologische Grundlagen und Technik der Pflanzen1 produktion. Zweite, neubearb. und erw. Aufl., Verlag Paul Parey, Berlin und Hamburg, pp 530. ISBN: 3-489-61510-7. https://doi.org/10.1002/jpln.19881510413
Hendl M (1963) Systematische Klimatologie – Einführung in die physikalische Klimatologie. Verlag der Wissenschaften, Berlin, II. VEB Dt
Holdridge LR (1947) Determination of World Plant formations from simple climatic data. Science 105(2727):367–368. https://doi.org/10.1126/science.105.2727.367
Holdridge LR (1959) Simple method for determining potential evapotranspiration from temperature data. Science 130(3375):572. https://doi.org/10.1126/science.130.3375.572.a
Holdridge LR (1967) Life zone ecology. Tropical Science Centre, San Jose. https://app.ingemmet.gob.pe/biblioteca/pdf/Amb-56.pdf
Humboldt AV, Bonpland A (1805) Essai sur la géographie des plantes accompagné d'un tableau physique des régions équinoxiales, fondé sur des mesures exécutées, depuis la dixième degré de latitude boréale jusqu'au dixième degre de latitude australe, pendant les années 1799, 1800, 1801, 1802 et 1803. Levraut Schoell et Co. Paris, France
Köppen W (1900) Versuch einer Klassifikation der Klimate, vorzugsweise nach ihren Beziehungen zur Pflanzenwelt. Geographische Zeitschrift, vol 6: 593–611 und 657–679
Köppen W (1923) Die Klimate der Erde – Grundriss der Klimakunde, Berlin, Leipzig
Köppen W (1936) Das geographische System der Klimate. Handbuch der Klimatologie. Gebrüder Bornträger, Berlin
Kottek M, Grieser M, Beck Ch, Rudolf B, Rubel F (2006) World map of the Köppen-Geiger climate classification updated. Meteorologische Zeitschrift 15:259–263
Kupfer E (1954) Entwurf einer Klimakarte auf genetischer Grundlage. Zeitschrift Für Erdkunde-unterricht 6:5–13
Larcher W (1980) Ökologie der Pflanzen. UTB, Stuttgart, p 232
Larcher W (1994) Ökologie der Pflanzen, 5th edn. Ulmer Verlag, Stuttgart, UTB für Wissenschaft

Lauer W (1952) Humide und aride Jahreszeiten in Afrika und Südamerika und ihre Beziehungen zu den Vegetationsgürteln. Bonner Geographische Abhandlungen 9:15–98

Lauer W (1981) Ecoclimatological Conditions of the Páramo belt in the Tropical High Mountains. Mt Res Dev 1(3–4):209–221

Lauer W (1991) Climatology. In: Pancel L (ed) Tropical forestry handbook. Springer Verlag, Berlin, Heidelberg, New York, pp 95–164

Lauer W, Frankenberg P (1985) Versuch einer geoökologischen Klassifikation der Klimate. Geogr. Rundschau, Jg. 37(7):359–365. Braunschweig https://www.adwmainz.de/wilhelm-lauer-stiftung/publikationen.html

Lauer W, Frankenberg P (1988) Klimaklassifikation der Erde. Geogr. Rundschau 40. Westermann Verlag, Braunschweig

Lauer W, Klaus D (1975a) Geoecological investigations of the timberline of Pico de Orizaba. Arctic Alpine Res 7:315–330

Lauer W, Klaus D (1975b) The thermal circulation of the central Mexican Meseta region within influence of the trade winds. Arch. für Meteorologie Geophysik und Bioklimatologie. Serie B 23:342–366

Lauer W, Rafiqpoor MD (2002) Die Klimate der Erde – eine Klassifikation auf der Grundlage der ökophysiologischen Merkmale der realen Vegetation. Franz Steiner Verlag, Stuttgart, p 271

Lauer W, Rafiqpoor MD, Frankenberg P (1996) Die Klimate der Erde – eine Klassifikation auf ökophysiologischer Grundlage der realen Vegetation. Erdkunde 50:257–300

Loidi J, Navarro-Sánchez G, Vynokurov D (2022) Climatic definitions of the world's terrestrial biomes. Veget Classification Surv 3:231–271. https://www.researchgate.net/publication/366408169_Climatic_definitions_of_the_world's_terrestrial_biomes

Lomolino MV, Riddle BR, Brown JH (2005) Biogeography, 3rd edition. Sinauer Associates, Sunderland. https://media-natur.com/Lomolino-Riddle-Brown-Biogeography

Lozan JL, Breckle S-W, Graßl H, Kasang D (eds) (2021) Warnsignal Klima: Boden und Landnutzung. Wissensch. Auswertungen, Geo, Hamburg. https://www.fdr.uni-hamburg.de/record/9905

Mucina L, Wasserman J (2026) Schimperian World: the foundations of biome ecology. Flora. https://doi.org/10.1016/j.flora.2026.152953

Neef E (1954) Allgemeine physische Erdkunde. Lehrbuch für das 9. Schuljahr, Berlin

Peel MC, Finlayson BL, McMahon TA (2007) Updated world map of the Köppen-Geiger climate classification. Hydrol Earth Syst Sci 11(5):1633–1644. https://doi.org/10.5194/hess-11-1633-2007

Rehm S, Espig G (1984) Die Kulturpflanzen der Tropen und Subtropen. Stuttgart

Ritter M (2003) The physical environment: an introduction to physical geography. In: Online textbook: learning environment, pp 450

Sanderson M (1999) The classification of climates from Pythagoras to Koeppen. Bulletin of Amer. Meteor. Soc. 80:669–673

Siegmund A, Frankenberg P (1999) Die Klimatypen der Erde – ein didaktisch begründeter Klassifikationsversuch. Geogr Rundsch 51(9):494–499

Stephenson NL (1990) Climatic control of vegetation distribution: the role of the water balance. Am Nat 135:649–670

Strahler AH, Strahler AN (2005) Physical geography: science and systems of the human environment. Wiley & Sons, New York, p 794

Strahler AN (1971) The earth sciences. Harper and Row, New York, pp 824

Thornthwaite CW (1943) Problems in the classification of climates. Geogr Rev 33(2):233–255

Thornthwaite CW (1948) An approach toward a rational classification of climate. Geogr Rev 38:55–94

Troll C (1964) Karte der Jahreszeitenklimate der Erde – mit einer farbigen Karte von Troll C und Paffen KH. Erdkunde 18:5–28. https://doi.org/10.3112/erdkunde.1964.01.02

Walter H, Breckle S-W (1983ff., 1986, 1991, 1994) Ökologie der Erde. Bd I, II, III, IV. Stuttgart

Walter H, Breckle S-W (1994): Ökologie der Erde Bd I, II, II, IV. Stuttgart

Walter H, Lieth, H (1960–67) Klimadiagramm-Weltatlas. VEB Gustav Fischer Verlag Jena
Whittaker RH (1975) Communities and ecosystems, 2nd edn. Macmillan, New York, p 385
Winiger M (1979) Bodentemperaturen und Niederschlag als Indikatoren einer klimaökologischen Gliederung tropischer Gebirgsräume. Methodische Aspekte und Anwendbarkeit dargelegt am Beispiel des Mt. Kenya. Geomethodica 4:121–150
Winiger M (1981) Zur thermisch-hygrischen Gliederung des Mount Kenya. Erdkunde 35:248–263
Wissmann HV (1939) Die Klima– und Vegetationsgebiete Eurasiens. Zeitschrift der Gesellschaft für Erdkunde zu Berlin: 1–14

Chapter 2
The Köppen's Climate Classification and Its Perception from 1900 to Today

2.1 Introduction

There is not enough space in this context to discuss the classification of Köppen (1900) in detail and to point out its weaknesses and strengths; reference is made here to Lauer (1999). A detailed discussion of the previously presented climate classifications (Köppen 1900, 1923, 1936; von Wissmann 1939; Holdridge 1947, 1959, 1967; Creutzburg 1950; Neef 1954; Hendl 1963; Troll 1964) etc. is deliberately omitted. It is only pointed out that the climate classification of Köppen shows inconsistencies with regard to the selection of boundary criteria (Trewartha 1943; Lauer 1975, 1991, 1999). While this classification thermally separates the different climate types by temperature thresholds, it uses mean precipitation and a temperature/precipitation-based "aridity index" as hygric parameters for the delimitation of arid climates (Lauer 1999; Lauer and Rafiqpoor 2002). Furthermore, Köppen lumps together all arid regions of the globe, regardless of their thermal characteristics, into a **B Climate**. For example, tropical and subtropical deserts and steppes are not distinguished (Lauer 1991, 1999; Lauer and Rafiqpoor 2002; Rafiqpoor 2024). It also neglects the solar climatic boundaries between the major climatic zones of the globe, with the result that areas with fundamentally different thermal characteristics are aggregated into inadequate climate types (Lauer 1975; Lauer and Frankenberg 1985, 1988; Lauer and Rafiqpoor 2002; Rafiqpoor 2024).

In his book "**Climatology**—Climate research in the twenty-first century—A challenge for the natural and social sciences", Martin Kappas emphasises that: "*In a textbook on climate research in the twenty-first century, the topic of 'climate classifications' initially seems antiquated. However, if you want to deal with the possible effects of climate change on the Earth's climate zones, it is very important to know the characteristics and geographical distribution of the average current state of the Earth's climate zones (e.g. assessment of possible climate projections such as the SRES projections: B1, A1B, A2* etc.)" (Kappas 2009: 142). That is precisely our aim

M. D. Rafiqpoor and S.-W. Breckle, *The World's Ecoclimates*,
SpringerBriefs in Earth System Sciences,
https://doi.org/10.1007/978-3-032-06456-1_2

in Part II of the book: to analyse the SRES scenarios as well as the climate zone shifts caused by climate change, using our new ecoclimate classification model.

One of the recent attempts is the digital climate classification of Guetter and Kutzbach (1990), which simulates past climates based on the modification of Köppen's climate zones according to Walter's and Breckle's (1983, 1984, 2004) zonobiome boundaries. The Köppen's classification is also used as a basis for landscape transformation as a result of climate change (Kalvova et al. 2003); for the assessment of global climate types during the Last Glacial Maximum and the present (Yoo and Rohli 2015); for global change time slices (Beck et al. 2023); or for diagnosing errors in Atmosphere–Ocean General Circulation Models relevant to the terrestrial biosphere (Gnanadesikan and Stouffer 2006).

The Köppen's (1900) classification, however, continues to form the basis for the development of a future climatology, which is implied by the output of a coarse resolution numerical climate model (Lohrmann et al. 1993) based on sea level temperature data averaged over the period 1951–1980 (Jones et al. 1991) and precipitation data for the period 1920–1980 (Legates and Willmott 1990). Fraedrich et al. (2001), using CRU-data (see New et al. 2002 and also https://crudata.uea.ac.uk/cru/data/hrg/) for the period 1901 to 1995, presented a Köppen map of 16 climate types at a resolution of 0.5° latitude by 0.5° longitude and examined the change in climate types over the period 1981 to 1995 relative to the full period of record. Development of a computer Programme Sparovek et al. (2007) automated the Köppen's classification to create a new climate map of Brazil and to compare the limits of their own classification with that of Köppen. Based on high-resolution data sets, Feddema (2005) provides Thornthwaite's (1948) classification as a digital cartographic map, which did not previously exist in this form.

However, the most comprehensive gridded Köppen's world map to date is that of Beck et al. (2023), which uses the Köppen-Geiger climate classification map for historical (based on high-resolution, observation-based climatologies) and future climate conditions (based on downscaled and bias-corrected climate projections for seven common socio-economic pathways, SSPs). In his digitized map, Kottek et al. (2006) differentiate the Earth into 31 climate types (Fig. 2.1) with a lat/long resolution of 0.5° × 0.5°, based on both CRU and GPCC (Global Precipitation Climatology Centre, German Weather Service: DWD) data sets for the period 1951–2000. Kottek et al. (2006) state here that "*while the climate classification concept of Wladimir* Köppen (1900), *presented in its latest version in 1961 by Rudolf Geiger, has been widely applied to a wide range of topics in climate and climate change research, as well as in physical geography, hydrology, agriculture, biology and educational aspects, a well-documented update of the world climate classification map is still lacking*". In this way, they wanted to present an "*updated digitized Köppen-Geiger world map for the second half of the twentieth century*" (Kottek et al. 2006). Peel et al. (2007) pointed out that the above-mentioned maps based on gridded data are for limited periods (1901–19,195, 1961–1990 or 1951–2000) and therefore any variability of climate types at sub-grid resolution is obscured. For this reason, they produced a new global climate map using the same Köppen-Geiger (1961) system, based on a large Global Historical Climatology Network (GHCN) dataset (Peterson and Vose 1997) of long-term monthly precipitation and temperature time series of climate stations for the entire period of record. In this paper, the climatic variables used in the Köppen's system were calculated at each station and interpolated between stations using a two-dimensional (lat/long) thin-plate spline with tension on a 0.1° × 0.1° grid separately for each continent.

In order to understand the Earth's climate, it is necessary to break down the vast amount of information into smaller areas of interest. In fact, it is quite obvious that the Earth's climates, as a continuum, cannot be demarcated by rigid lines. At best, there are transitional areas between climatic regions, as there are between vegetation types. But the question is, "**why do we always draw boundaries?**". We need them to understand something effectively and easily!

A boundary itself always acts as a regulatory factor in human life, although, with the exception of the boundary between land and water or ice (snow), as well as some 'line-like' (e.g., streams, road networks) or 'point-like' elements (e.g., bogs, wetlands, swamps), there are rarely sharp limits in nature, i.e., natural phenomena are always continuous. In most cases, a boundary is not an abrupt break; the different phenomena will gradually merge into each other, resulting in real transition zones as "**boundary-strips**" between regions, instead of the rigid (cultural or natural) boundaries created by humans (Leimgruber 1994). Those transition zones are called **ecotones** or **zonoecotones** by Walter and Lieth (1960–67) and Breckle (2002).

Scientists are always striving to make natural phenomena tangible by finding the "common" in natural spatial units through delimitation. "*Behind the boundary there is ultimately a principle of order (a system): Regionalisation, whether physical or cultural, emphasises the common over the divisive at various levels of scale, thus combining similar (spatial) phenomena into superordinate (spatial) units*" (Leimgruber 1994).

All climate classifications are therefore concerned with reconstructing the optimal boundaries for regions with the same climatic characteristics (**'homoclimates'**). For this reason, criteria and thresholds corresponding to certain phenomena in nature are used as a kind of **'ground check'**. For example, Köppen (1900) uses the global vegetation zones of De Candolle (1874) (see above) to validate climate types with natural units. Ecological climatic diagrams according to Walter and Lieth (1960–67) (see Fig. 4.10) are particularly useful to trace such **homoclimates**, i.e. stations with similar or almost identical climates (see Fig. 4.12). This can be a tedious process if you have to consult lengthy climatic tables with meteorological data, but ecological climate diagrams make it easy to compare a given diagram with regions where homoclimates are suspected.

The shortcomings of Köppen's (1900) classification, i.e. the neglect of the solar-climatic boundaries of the main climate zones or the aggregation of arid climates, also apply to the classifications of Kottek et al. (2006) and Peel et al. (2007). The latter classify, for example, the climate station Herberton Post Office (Fig. 2.2) (17.38° S, 145.38° E, 899 m asl, Queensland, Australia) in the **Cw** climate type of Köppen, although this station, with about 1146 mm annual precipitation and an annual temperature range of the monthly means of only 7.7 K, is definitely a tropical savannah climate type in the elevational belt of lower "*tierra templada*" (Lauer 1975; Lauer

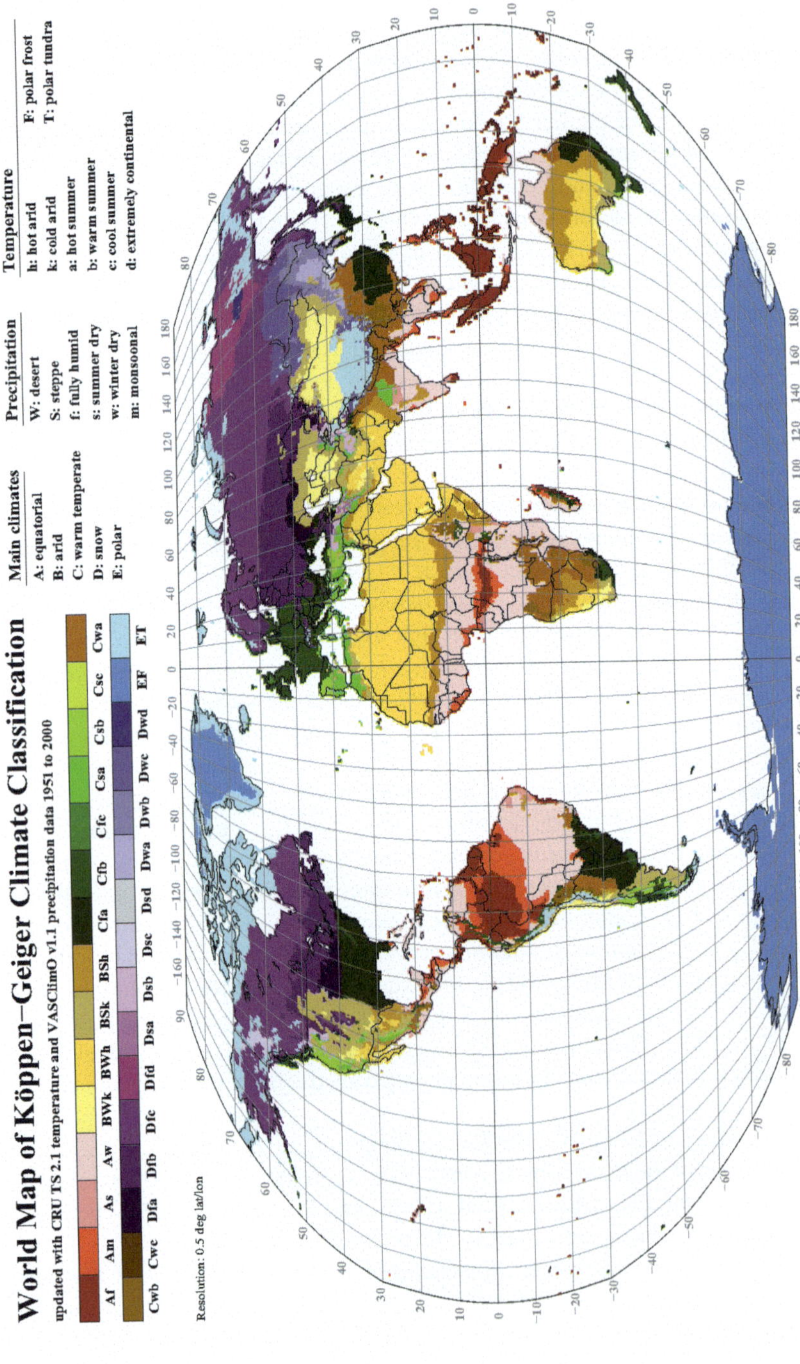

Fig. 2.1 The Climate Classification of Köppen (1900) according to attempts by Kottek et al. (2006)

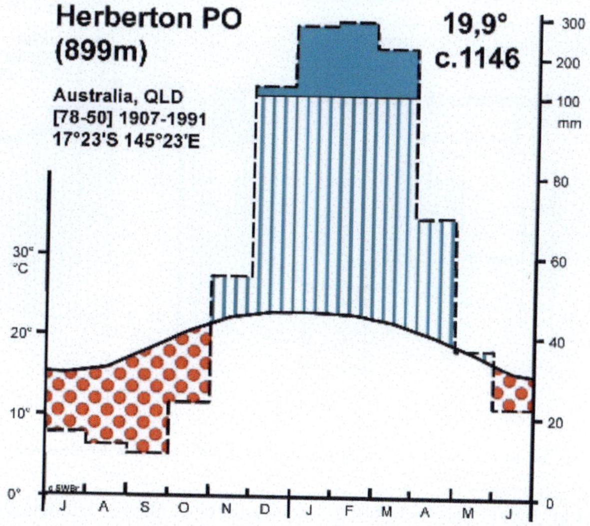

Fig. 2.2 Ecological climate diagram of Herberton Post Office in Queensland, Australia (details to ecological climatic diagrams see p. 55, Fig. 4.11 and p. 56ff.)

and Rafiqpoor 2002), where in the long-term average, due to sporadic cold air break-ins in winter, absolute minimum temperatures of +13 °C can occasionally occur. However, this does not mean that this region can be assigned to a **Cw** climate type of the outer-tropics.

Looking at the different continental maps of Peel et al. (2007), it is striking that in Africa, for example, the Ethiopian highlands or the entire dry savannah area north of the Zambezi River up to the vicinity of the Congo Basin are assigned to the (subtropical) **Cs** and **Cw** climate types. On the continental map of Asia, the temperate **Cw** subtropical climate also extends from southern Japan, along the marginal tropics of southern China, down the northern slopes of the Himalayas to the Ganges lowlands of northern and central India. Any researcher familiar with the nature of South America will wonder why the **Cw** climate is already assigned to the **Chocó region** as a global centre of biodiversity (Barthlott et al. 2014) on the equatorial Pacific coast of Ecuador and southern Colombia. Peel et al. (2007), for example, explain the distribution of the **Cw** climate around Nairobi ("*equatorial Kenya*" see as type-diagram Mombasa Fig. 2.3, left) or around Antananarivo (Madagascar; see as type-diagram Toamasina, Fig. 2.3, right) by the climate stations in the adjacent lowland regions.

The digital map of Köppen by Kottek et al. (2006) (see Fig. 2.1) also shows the same shortcomings. In this classification within the tropics, an extratropical **Cw** climate is assigned, for example, to the monsoon-influenced western slopes of the Himalayas down to the lowlands of the Ganges and Brahmaputra, as well as to the above-mentioned regions in southern China (see also Peel et al. 2007 and Fig. 2.1). With regard to the spatial extent of the tropical rainforest climate, the two latter classifications are also not sufficiently informative. For example, large parts of the rainforest climate in the Amazon basin of South America are classified as tropical

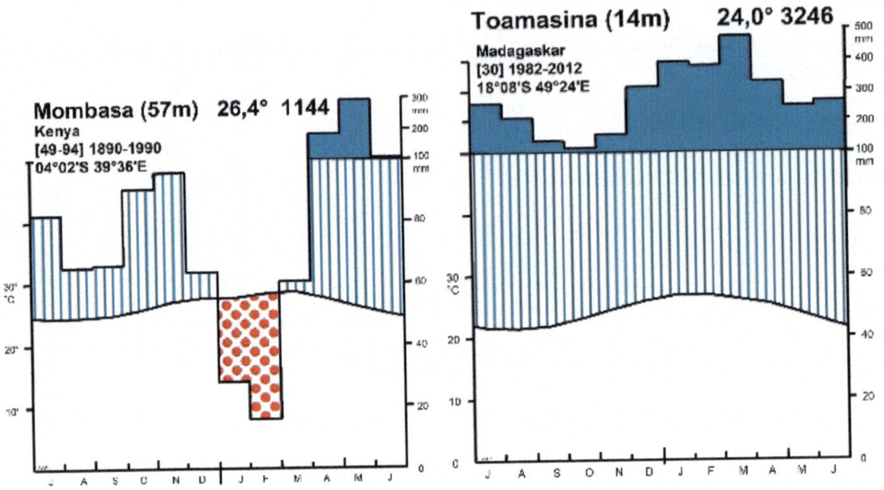

Fig. 2.3 Ecological climate diagrams of Mombasa (Equatorial Africa) and Toamasina (Madagascar) as typical tropical lowland stations exhibiting a small yearly temperature range

savannah. Especially in the classification of Kottek et al. (2006) this phenomenon is much more pronounced.

The new climate classification of Australia (Stern et al. 2000) attempts to regroup subdivisions of the Köppen's climate types. Even in this classification, the use of 'average climatology' data actually obscures the climatic characteristics of many regions. For example, the climate of the northern tip of Queensland is characterised as equatorial, even though this region has a very pronounced seasonal rainfall regime (Beadle 1981). Although in some orographically favourable regions, even on the east coast of Queensland, annual precipitation values of more than 2000 mm (i.e. in Cairns 2075 mm, in Innisfail 3563 mm, both in Queensland: Fig. 2.4) can fall with lush rainforest vegetation as a result of precipitation caused by trade winds on the windward side of the coasts, the area cannot be classified as an equatorial climate type, but rather as an area with an alternating tropical megatherm-subhumid climate (savannah climate according to Troll 1964).

Even more misleading is in this classification the boundary between tropical and subtropical climates. Also, attempts to divide **temperate climates** into two groups (i.e. subtropical and temperate) are not always successful because the thermal conditions (differentiation between tropical/subtropical and subtropical/temperate temperature regimes) and the solar climatic boundaries of the macroclimatic zones are neglected.

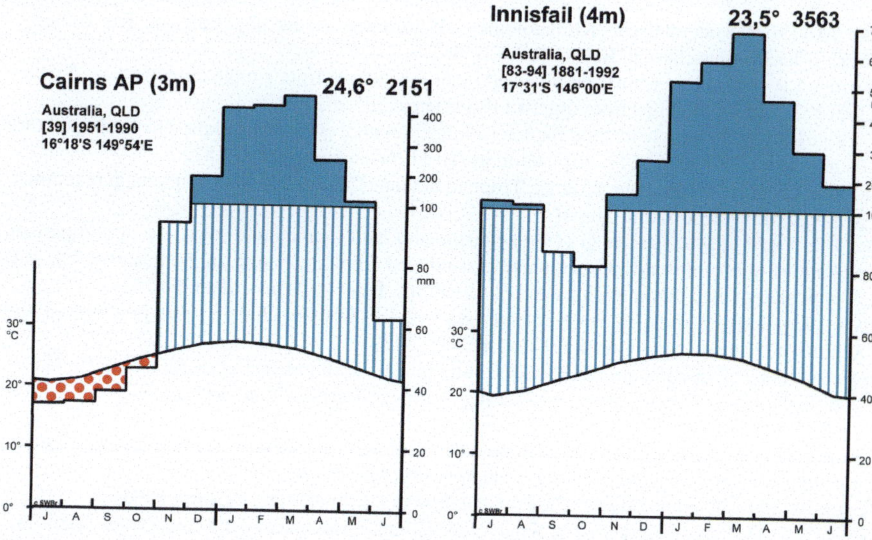

Fig. 2.4 Ecological climate diagrams of Cairns and Innisfall, two tropical stations on the east coast of Queensland, Australia, with upslope rainfall caused by trade winds and lash-green tropical rainforest vegetation

References

Barthlott W, Erdelen RW & Rafiqpoor MD (2014) Biodiversity and technical innovations: bionics. In: Lanzerath D, Friele M (eds) Concept and Value in Biodiversity. Routledge Studies in Biodiversity Politics and Management. pp 300–315. ISBN: 978-1-415-66057-0. https://www.routledge.com/Concepts-and-Values-in-Biodiversity/Lanzerath-Friele/p/book/9780415660570

Beadle NCW (1981) The vegetation of Australia. In: Walter H, Breckle S-W (eds) Vegetation-Monographien der einzelnen Großräume, vol IV, Fischer, Stuttgart pp 690

Beck HE, McVicar TR, Vergopolan N, Berg A, Lutsko NJ, et al. (2023) High-resolution (1 km) Köppen-Geiger maps for 1901–2099 based on constrained CMIP6 projections. Scientific Data 10, Article number: 724, pp 16. https://doi.org/10.1038/s41597-023-02549-6

Breckle S-W (2002) Walter's Vegetation of the earth. The ecological systems of the Geo–Biosphere. Springer, Berlin

Creutzburg N (1950) Klima, Klimatypen und Klimakarten. Petermanns Geogr Mitt 49:59–69

De Candolle A (1874) Constitution dans le végétal de groups physiologiques applicable á la géographie botanique ancienne et modern. Univers, Arch. d. Scien. Bibl

Feddema JJ (2005) A revised thornthwaite-type global climate classification. Phys Geogr 26(6):442–466. https://doi.org/10.2747/0272-3646.26.6.442

Fraedrich K, Gerstengarbe FW, Werner PC (2001) Climate shifts during the last century. Clim Change 50:405–417. https://doi.org/10.1023/A:1010699428863

Gnanadesikan A, Stouffer RJ (2006) Diagnosing atmosphere-ocean general circulation model errors relevant to the terrestrial biosphere using the Köppen climate classification. Geogr Res Lett 33:1–5

Guette PJ, Kutzbach JE (1990) A modified Köppen classification applied to model simulations of glacial and interglacial climates. Clim Change 16:193–215. https://doi.org/10.1007/BF00134657

Hendl M (1963) Systematische Klimatologie – Einführung in die physikalische Klimatologie. Verlag der Wissenschaften, Berlin, II. VEB Dt

Holdridge LR (1947) Determination of world plant formations from simple climatic data. Science 105(2727):367–368. https://doi.org/10.1126/science.105.2727.367

Holdridge LR (1959) Simple method for determining potential evapotranspiration from temperature data. Science 130(3375):572. https://doi.org/10.1126/science.130.3375.572.a

Holdridge LR (1967): Life zone ecology. Tropical Science Centre, San Jose. https://app.ingemmet.gob.pe/biblioteca/pdf/Amb-56.pdf

Jones PD, Vigley TML, Farmer G (1991) Marine and land temperature data sets. A comparison and a look at recent trends. Greenhouse gas-induced climatic change. In: Schlesinger ME (ed) A critical appraisal of simulations and observations. Elsevier, Amsterdam, pp 153–172

Kalvova J, Halenka T, Bezpalcova K, Nemesova I (2003) Köppen climate types in observed and simulated climates. Stud Geophys Geod 47:185–202

Kappas M (2009) Klimatologie – Klimaforschung im 21. Jahrhundert. Herausforderung für Natur- und Sozialwissenschaften. Spektrum Verlag, Springer, Berlin, Heidelberg, pp 542. ISBN: 978-3-662-62105-9

Köppen W (1900) Versuch einer Klassifikation der Klimate, vorzugsweise nach ihren Beziehungen zur Pflanzenwelt. Geographische Zeitschrift 6:593–611 und 657–679

Köppen W (1923) Die Klimate der Erde – Grundriss der Klimakunde, Berlin, Leipzig

Köppen W (1936) Das geographische System der Klimate. Handbuch der Klimatologie. Gebrüder Bornträger, Berlin

Kottek M, Grieser M, Beck Ch, Rudolf B, Rubel F (2006) World map of the Köppen-Geiger climate classification updated. Meteorologische Zeitschrift 15:259–263

Lauer W (1991) Climatology. In: Pancel L (ed) Tropical forestry handbook. Springer Verlag, Berlin, Heidelberg, New York, pp 95–164

Lauer W, Rafiqpoor MD (2002) Die Klimate der Erde – eine Klassifikation auf der Grundlage der ökophysiologischen Merkmale der realen Vegetation. Franz Steiner Verlag, Stuttgart, p 271

Lauer W, Frankenberg P (1985) Versuch einer geoökologischen Klassifikation der Klimate. Geogr. Rundschau, Jg. 37(7):359–365. Braunschweig https://www.adwmainz.de/wilhelm-lauer-stiftung/publikationen.html

Lauer W, Frankenberg P (1988) Klimaklassifikation der Erde. Geogr. Rundschau 40. Westermann Verlag, Braunschweig

Lauer W (1975) Vom Wesen der Tropen – klimaökologische Studien zum Inhalt und zur Abgrenzung eines irdischen Landschaftsgürtels. Abhandlungen der Math.-nat. Klasse, Akademie d. Wissenschaften und der Literatur, Mainz, Jg. 3: 52 pp. https://www.adwmainz.de/wilhelm-lauer-stiftung/publikationen.html

Lauer W (1999) Klimatologie. Das Geographische Seminar – Westermann-Verlag, Braunschweig, pp 270. ISBN: 3-14-160284-0

Legates DR, Willmott CJ (1990) Mean seasonal and spatial variability in gauge-corrected, global precipitation. Int J Climatol 10:111–127

Leimgruber W (1994) Was ist eine Grenze? In: Universitas Friburgensis avril 1994, Fribourg, Swiss

Lohrmann U, Sausen R, Bengtsson L, Cubasch U, Perlwitz J et al (1993) The Köppen climate classification as a diagnostic tool for general circulation models. Climate Res 3:177–193. https://doi.org/10.3354/cr003177

Neef E (1954) Allgemeine physische Erdkunde. Lehrbuch für das 9. Schuljahr, Berlin

New M, Lister D, Hulme M, Makin I (2002) A high-resolution data set of surface climate over global land areas. Climate Res 21:1–25

Peel MC, Finlayson BL, McMahon TA (2007) Updated world map of the Köppen-Geiger climate classification. Hydrol Earth Syst Sci 11(5):1633–1644. https://doi.org/10.5194/hess-11-1633-2007

Peterson TC, Vose RS (1997) An overview of the global historical climatology network temperature database. Bull Am Meteorol Soc 78:2837–2849

Rafiqpoor MD (2024): Climatology اقليم شناسی—DCM-Druck Center Meckenheim. Germany, pp 298. https://www.researchgate.net/publication/377066161_Climatology_-_aqlym_shnasy

Sparovek G, Van Lierb Q, De J, Neto DD (2007) Computer assisted Koeppen climate classification: a case study for Brazil. Intern J Climatol 27(2):257–266. https://doi.org/10.1002/joc.1384

Stern H, De Hoedt G, Ernst J (2000) Objective classification of Australian climate. Aust Meteor Mag 49(2):87–96. https://www.researchgate.net/publication/279766565_Objective_classification_of_Australian_climates

Thornthwaite CW (1948) An approach toward a rational classification of climate. Geogr Rev 38:55–94

Trewartha GT (1943) An introduction to weather and climate. McGraw-Hill, New York, p 545

Troll C (1964) Karte der Jahreszeitenklimate der Erde – mit einer farbigen Karte von Troll C und Paffen KH. Erdkunde 18:5–28. https://doi.org/10.3112/erdkunde.1964.01.02

Walter H, Breckle S-W (1983, 1984, 2004) Ökologie der Erde Bd. 2: Spezielle Ökologie der tropischen und subtropischen Zonen. Stuttgart

Walter H, Lieth H (1960–67): Klimadiagramm-Weltatlas. VEB Gustav Fischer Verlag Jena

Wissmann H v (1939) Die Klima- und Vegetationsgebiete Eurasiens. Zeitschrift der Gesellschaft für Erdkunde zu Berlin, 1–14

Yoo J, Rohli RV (2015) Global distribution of Köppen-Geiger climate types during the Last Glacial Maximum, Mid-Holocene, and present. Palaeogeography Palaeoclimatology Palaeoecology 446. https://doi.org/10.1016/j.palaeo.2015.12.010

Chapter 3
Why Do We Need a New Digital Ecoclimate Classification?

In view of the above-mentioned shortcomings of existing attempts to digitise the Köppen-Geiger (1961) climate classification, many authors complain that no quantitatively verifiable ecological climate classification based on large data sets has yet been successfully developed. For example, "*the more than 100-year-old climate classification of Wladimir* Köppen (1900) *is still regularly used in school and undergraduate courses on climate*" (Peel et al. 2007), as well as by researchers in various disciplines as a basis for climatic regionalisation of variables and for assessing the output of global climate models (Kalvova et al. 2003; Beck et al. 2005, 2023), but also for practical purposes (Briggs et al. 2002).

The question "**Why do we need a new digital ecoclimate classification**?"—even at the cost of accusations of antiquity (Kappas 2009)—is now more obvious than ever, as Walter Erdelen (https://wsds.teriin.org/2010/bio2010/walter_erdelen.pdf) rightly noted in a personal communication: "*In the current debate on climate change, the climate itself has really been forgotten*".

It has been verified for certain areas of the world that the Köppen's system reacts rather insensitively to the observed global warming, and that the system performs rather poorly overall, especially over Europe and Asia (Triantafyllou and Tsonis 1994). Knoch and Schulze (1954) have already discussed the problem of climate classification with regard to the continued applicability and the claim to universal validity and authority of Köppen's climate classification: "*The creation of additional classifications probably only makes sense if a large collection of as many important characteristic climatic variables as possible is achieved*" (Knoch and Schulze 1954).

With this in mind, Lauer and Rafiqpoor (2002) emphasised: "*The authors of the present ecoclimate classification believe that with improved data availability, the use of modern methods, and targeted objectives, climate in all its complexity can now be more specifically defined. The existing classifications, which are based on "average climatology", do not meet today's requirements. It therefore seems worthwhile to develop an ecologically designed climatic classification based on data on the* ***length of the thermal and hygric growing seasons*** *as parameters for the typification of space*

M. D. Rafiqpoor and S.-W. Breckle, *The World's Ecoclimates*,
SpringerBriefs in Earth System Sciences,
https://doi.org/10.1007/978-3-032-06456-1_3

as physiologically based heat and water balances" (see also Flohn 1957). This means that we can use publicly available data on monthly temperature and precipitation averages to determine the length of the thermal and hygric growing seasons (in months), which are basic parameters for the spatial typification of ecoclimates.

This brings us to the crux of the matter, and we join Sanderson in asking: "*Modern atlases and geography textbooks continue to use the 100-year-old Köppen classification of climates, based on De Candolle's vegetation groups, which were themselves based on the five climatic zones of the ancient Greeks! Is it not time for modern atmospheric scientists to develop a* ***new*** *('digital') classification of world climates?*" (Sanderson 1999). Even if an attempt is made to create a new digital ecoclimate classification based on the variation in day length defined by the radiation climate, the proven five traditional climate zones will be retained.

Given the existing global data sets, Bailey (author of "Ecoregions—The Ecosystem Geography of the Oceans and Continents"), in his discussion paper "Ecological Climate Classification—It's more than the mean monthly temperature of January or the mean annual precipitation", insists on the need for an ecological climate classification: "*Large amounts of climate data exist in databases or as data layers in geographic information systems. Such raw data have not yet been synthesised into* **ecological climate classes** *that are useful for ecological analysis*" (Bailey 2014). Bailey probably refers to a **digital climate classification**. With the help of 8000 climate diagrams, Walter (1954), Walter and Lieth (1960) made a detailed classification of the earth into nine zonobiomes and intermediate zonoecotones (Walter and Breckle 1991, 2004; Breckle 2021).

In their semi-quantitative climate classification, Lauer and Rafiqpoor (2002) have already attempted to match the distribution limits of climate types to the relief and vegetation formations. As this classification approach uses the natural and the real vegetation (see Fig. 1.2) as a reference, it takes into account the quantitatively measurable hygrothermal parameters of the ecological requirements of the vegetation in its annual rhythm. Box (2004) in his review of this classification system states: "*Wilhelm Lauer was perhaps first known (1952) for his demonstration that it was not the total amount of annual precipitation but rather the length of the wet and dry seasons that best correlated with the natural landscape in the tropics. … The system presented by* Lauer and Rafiqpoor (2002) *is now quite thought-provoking, the theoretical and empirical basis for quantification is useful and well executed, and the description of different world regions is often very good*" (Box 2004). In this context, Leser emphasises that "*the developed classification represents by* Lauer and Rafiqpoor (2002) *is a fundamental advance in the classification of climates. It also has a forward-looking character in that the system "***Climate-Earth Surface-Vegetation***" will also change as a result of increasing human use of the Earth's surface and the quantitative justification of the boundaries of the climate types—with changes in the parameters—will make it possible to recalculate the areas*" (Leser 2005).

Thus, our contribution now aims to meet the demand for an ecologically oriented climate classification in which climate types are quantitatively determined by their specific hygrothermal properties, similar to the indication of relative arid (red stippled) and relative humid (blue hatched) months in the ecological climatic diagrams (see Fig. 4.11). We also provide here a classification system that has the potential to predict future climates based on climate change scenarios (see Part II).

References

Bailey RG (2014) Ecological climate classification. In: Ecoregions – The ecosystem geography of the oceans and continents. Springer. https://doi.org/10.1007/978-1-4939-0524-9

Beck C, Grieser J, Kottek M, Rubel F, Rudolf B (2005) Characterizing global climate change by means of Köppen climate classification. In: Klimastatusbericht: 139–149. Deutscher Wetterdienst, Offenbach. https://www.researchgate.net/publication/255532038

Beck HE, McVicar TR, Vergopolan N, Berg A, Lutsko NJ et al (2023) High-resolution (1 km) Köppen-Geiger maps for 1901–2099 based on constrained CMIP6 projections. Scientific Data 10, Article number: 724, pp 16. https://doi.org/10.1038/s41597-023-02549-6

Box EO (2004) Review to the book of Lauer W & Rafiqpoor MD (2002) Die Klimate der Erde: Eine Klassifikation auf der Grundlage der ökophysiologischen Merkmale der realen Vegetation. Phytocoenologia 34(4):673–675. https://www.schweizerbart.de/papers/phyto/list/34#issue1

Breckle S-W (ed) (2021) Ökologie der Erde. Band 3. Spezielle Ökologie der Gemäßigten und Arktischen Zonen Euro-Nordasiens. Zonobiom VI–IX. 3. Aufl., Schweizerbarth, Stuttgart

Briggs RS, Lucas RG, Taylor ZT (2002) Climate classification for building energy codes and standards. In: Technical paper final review draft. 34 pp. Pacific NW National laboratory

Flohn H (1957) Zur Frage der Einteilung der Klimazonen. Erdkunde 11:161–175

Kalvova J, Halenka T, Bezpalcova K, Nemesova I (2003) Köppen climate types in observed and simulated climates. Stud Geophys Geod 47:185–202

Kappas M (2009) Klimatologie – Klimaforschung im 21. Jahrhundert. Herausforderung für Natur- und Sozialwissenschaften. Spektrum Verlag, Springer, Berlin, Heidelberg, pp 542. ISBN: 978-3-662-62105-9

Knoch K, Schulze A (1954) Methoden der Klimaklassifikation. Petermanns Geographische Mitteilungen Ergänzungsheft 249: 78 pp

Köppen W (1900) Versuch einer Klassifikation der Klimate, vorzugsweise nach ihren Beziehungen zur Pflanzenwelt. Geographische Zeitschrift, vol 6: 593–611 und 657–679

Lauer W, Rafiqpoor MD (2002) Die Klimate der Erde – eine Klassifikation auf der Grundlage der ökophysiologischen Merkmale der realen Vegetation. Franz Steiner Verlag, Stuttgart, p 271

Leser H (2005) Review to the book of Lauer W. & Rafiqpoor MD "Die Klimate der Erde - Eine Klassifikation auf der Grundlage der ökophysiologischen Merkmale der realen Vegetation". Geographica Helvetica 60:68–69

Peel MC, Finlayson BL, McMahon TA (2007) Updated world map of the Köppen-Geiger climate classification. Hydrol Earth Syst Sci 11(5):1633–1644. https://doi.org/10.5194/hess-11-1633-2007

Sanderson M (1999) The classification of climates from Pythagoras to Koeppen. Bull Amer Meteor Soc 80:669–673

Triantafyllou GN, Tsonis AA (1994) Assessing the ability of the Köppen system to delineate the general world pattern of climates. Geophys Res Lett 21(25):2809–2812

Walter H (1954) Einführung in die Phytologie. Band III: Grundlage der Pflanzenverbreitung, Teil I: Standortlehre, Stuttgart

Walter H, Breckle S-W (1991) Ökologie der Erde Band 1 – Grundlagen. 2. Auflage. Fischer Verlag, Stuttgart

Walter H, Breckle S-W (2004) Ökologie der Erde Band 2 – Spezielle Ökologie der tropischen und subtropischen Zonen. 3. Aufl. Spektrum Akad. Verl., München

Walter H, Lieth H (1960–67): Klimadiagramm-Weltatlas. VEB Gustav Fischer Verlag Jena

Chapter 4
Material and Method (Data)—The Basic Principles of Ecoclimate Classification

To develop an ecologically oriented climate classification, we used the extensive high-resolution WorldClim dataset (Hijmans et al. 2005; see https://www.worldclim.org/data/worldclim21.html) for temperature and precipitation, and the FAO dataset for potential evapotranspiration (PET) (Zomer and Trabucco 2022: The PET data are available at: https://doi.org/10.6084/m9.figshare.20005589), calculated according to the Penman and Monteith equation (Allen et al. 1998, Chap. 2: FAO Penman and Monteith equation; see http://www.fao.org/docrep/X0490E/x0490e06.htm).

From these datasets we calculated heat (length of thermal growing season in months—expressed in isothermomenes) and water balance (length of hygric growing season in months—expressed in isohygromenes) as the two main parameters of our ecological climate classification system. Due to the different spatial resolution of the WorldClim (30 arcSeconds) and FAO (30 arcMinutes) datasets, PET was interpolated using the SPLINE command of ArcGIS to create datasets of the same resolution (30 arcSeconds) for further calculations.

The concept of ecoclimate classification (Lauer et al. 1996; Lauer and Rafiqpoor 2002) is based not only on the natural but also on the **real vegetation cover** of the Earth as **indicator** of climate, since in today's world, the vegetation cover has changed almost extensively due to **land use** (Lozan et al. 2021). It is therefore meaningful to use also the real vegetation cover, as Lauer and Rafiqpoor (2002) have done, as the basis for evaluating the calculated climate types for an ecological climate classification. But in this contribution, to calculate the water budget, we used the precipitation data of WorldClim and the ETP data from FAO, calculated according to the Penman and Monteith method (https://www.fao.org/4/X0490E/x0490e06.htm), and not the pLV-data from Lauer and Rafiqpoor (2002), as the latter are not available in digital form. Therefore, the ecoclimate classification presented here is not only based on the real, but on the potential natural vegetation of the earth.

The vegetation cover at all needs, however, the heat balance for its life cycle and it controls the water and heat balance through its transpiration behaviour. It modifies the water balance through transpiration and the heat balance through sensible (V) and

M. D. Rafiqpoor and S.-W. Breckle, *The World's Ecoclimates*,
SpringerBriefs in Earth System Sciences,
https://doi.org/10.1007/978-3-032-06456-1_4

latent (L) heat fluxes. Thus, the higher the transpiration capacity of the vegetation cover, the greater the shift in the Bowen ratio (= L/V) in favour of the latent heat flux (Lauer and Frankenberg 1981a, 1982). If we cut down a forest, for example, we change the land surface and thus also the albedo and the evapotranspiration capacity of the landscape at that location. Therefore, any change in vegetation cover will affect climate, or conversely, any change in climate will affect vegetation cover.

An interesting example is demonstrated by the long-term Solling Project in Central Germany (Ellenberg et al. 1986). The annual net primary production of the sample plots, comprising beech forests, spruce forests, golden oat meadows, and ryegrass fields, was approximately the same throughout the year.

A corn field this year, a rapeseed field the next, and then perhaps the reforestation of the cultivated area obviously changes the climate little, but this depends heavily on the area size; the cultivated prairie in the Midwest of the USA or the steppes in southern Russia, as well as the large-scale deforestation of the tropical forests in the Amazon, all have an impact on the regional climate.

In this context, the amount of heat energy from the radiation balance at the Earth's surface is crucial for the ecophysiological processes in the vegetation cover. Temperature, as sensible heat, is also a component of the radiation balance. Through the radiation turnover at the Earth's surface, the temperature is significantly controlled by the vegetation cover (Kessler 1985). Since there is a constant exchange between vegetation cover and climate; **vegetation** is the best **indicator** of climate. For this reason, in the present ecoclimate classification, the **heat and water budgets** (or balances) are considered as the two basic **parameters** for the differentiation of the Earth's ecoclimate types, as is shown also by the shape of the temperature and precipitation curves in the ecological climate diagrams of Walter and Lieth (1960–67).

4.1 The Thermal Component of Ecoclimate Classification: Isothermomenes

Various measures have been used to describe the thermal component in ecoclimate classifications. Particularly popular are annual mean values of temperature, bio-temperature (as the annual average of all air temperatures above the freezing point. The bio-temperature is used to make comparative statements about the climatic growth conditions of vegetation) or potential evapotranspiration (cf. Holdridge 1947, 1959; Woodward and Lomas 2004; Woodward et al. 2004; Fisher et al. 2010). Indeed, all these energy-related variables are rather strongly colinear (Fig. 4.1).

The **thermal component** of the ecoclimate classification is defined as **isothermomenes** (Greek: ***iso*** = equal; ***thermos*** = heat, ***mene*** = month) as the **"monthly length of the thermal growing season"**.

Walter (1954) defines the length of the thermal growing season in the outer-tropics as the frost-free period of the year exceeding certain temperature thresholds. If data are available, those are marked in the ecological climate diagrams by months with

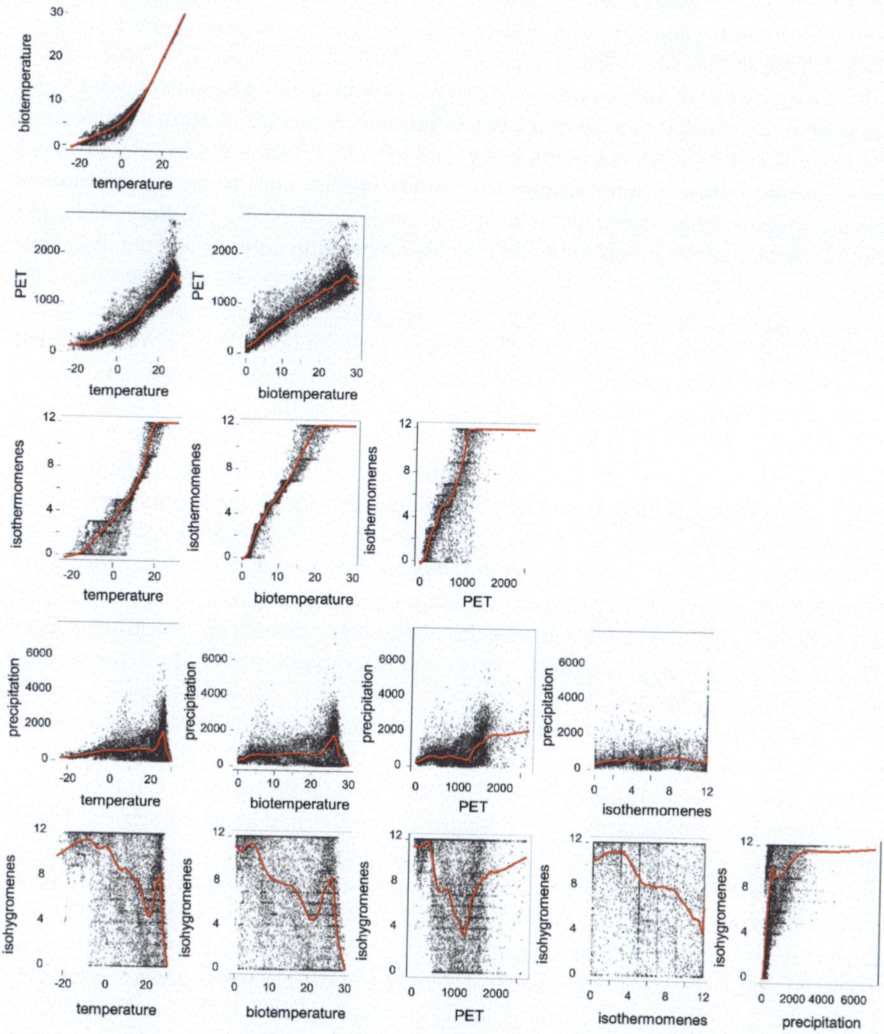

Fig. 4.1 Correlation and collinearity of frequently used climatic variables in different ecoclimatic classifications. Values were averaged across an equal area grid of 110 x 110 km. Lines represent locally weighted sums of squares trend lines (LOWESS; tension = 0.1). [**Units:** Isothermomenes and isohygromenes = months; Precipitation and PET = mm; Temperature and Bio-temperature = °C]

periodic or those with episodic frosts, or additionally with the number of days per year above 10 °C.

In terrestrial high-latitude ecosystems, the growing season begins in spring with increasing temperatures and light availability, melting of snow, thawing of soil organic horizons and the onset of photosynthesis. In autumn, the growing season ends as temperatures and light availability decrease, soils refreeze and photosynthesis ceases (Euskirchen et al. 2006).

Plants can extract a limited amount of resources from their environment, which can be allocated to growth either above or below ground. A species' pattern of allocation is genetically determined, but is not completely fixed (Johnson 2006). The growing season, during which in midlatitudes the woody species start to allocate resources generally begins when daily mean temperatures exceed 10 °C (Walter 1954). On the other hand, leaf discolouration and abscission begin in autumn when daily mean temperatures fall below about 10 °C. Although thermal thresholds for the onset and end of the generative phase may vary between different vegetation formations, a monthly mean temperature of ≥10 °C can generally be considered a robust global threshold. For example, the transition between taiga and tundra vegetation is demarcated when the length of the thermal growing season, i.e. the number of months ≥10 °C, falls below three (Walter and Breckle 1991). For many megathermal tropical plants, however, the 10 °C threshold marks the minimum cold tolerance (Hutchinson et al. 1992). Körner pointed out that "*if one disregards local/regional (including anthropogenic) reasons for the absence of trees, the high-altitude treeline is found at surprisingly uniform growing season temperatures worldwide, while other types of thresholds (including the warmest month temperature) do not work on a global scale*" (Körner 2007). At the treeline in the Alps, daily mean air temperature exceeds 5 °C for about 100 days (due to the existence of conifers at the alpine timberline) (more details see below).

Many authors define the thermal growing season as the number of days with an average daily temperature above 5 °C. In Europe, e.g. by the "UK Department for Environment, Food and Rural Affairs" (Mitchell and Hulme 2002), 5 °C is recognised as the critical threshold defining the length of the thermal growing season. Miller and Mitchel (2005) pointed out that the "US Department for National Agricultural Statistics Service" currently defines the growing season as the period between the last and first freezing temperatures (0 °C). Kunkel et al. (2004) also used this criterion. Many previous studies, e.g. Brown (1976), Moran and Morgan (1977), Brinkman (1979), Sharratt (1992), Feng and Hu (2004) used different temperatures ranging from −3 °C to +10 °C to determine the beginning and end of the growing season. But all these thermal thresholds are mainly related to crops and each crop reacts a bit different.

In order to depict an ecological climate classification on a global scale, it is not the thermal thresholds of crops or individual plants that can be used as a boundary criterion for climate types, but rather the geographical pattern of large-scale vegetation formations. For the onset of the thermal growing season, the greening of the entire tree vegetation is crucial, as it is only then that the photosynthetic process in the landscape unfolds at its full intensity (Walter 1954). On the other hand, it is known that the vegetation of the tundra and the boreal coniferous forests already breaks its winter dormancy in the spring, when the mean monthly temperature reaches 5 °C (Walter and Breckle 1991), and also ends its generative phase in the autumn, when the monthly mean falls below this critical level (Tuhkanen 1984; Treter 1993). Nevertheless, under "**long-day solar-climatic conditions**" in summer, tundra and boreal coniferous forest vegetation require temperatures of ≥10 °C for at least 1–2 and 3–4 months, respectively, to complete their generative phase. The thermal threshold for resource allocation in mid-latitude mixed forests is also 10 °C, for Mediterranean sclerophyllous trees and shrubs 12 °C, and for lowland tropical rainforests at least 18 °C (Lauer and Rafiqpoor 2002) (for a more detailed description, see page 46, Table 4.1).

The optimal temperature range between the beginning and end of the growing season is usually no wider than 10–25 °C for most plants (Larcher 1980) (for a more detailed description, see page 46, Table 4.1). For the global representation, however, we defined the length of the thermal growing season as the **number of months with mean monthly temperatures ≥10 °C** as a threshold for the minimum thermal requirement of **tree vegetation** (Walter 1954).

However, new approaches to a global treeline comparison by Körner (2021) have shown that the two completely independent assessments used provide strong support for his global treeline isotherm concept and thus for the global delineation of the lower edge of the alpine belt in the mountains. By combining nearly 400 treeline reference points from satellite imagery with the **WorldClim** data for site temperature, Paulsen and Körner (2014) arrived at a global treeline isotherm for a seasonal mean temperature of 6.1 °C. The smallest prediction error was a combination of the 6.1 °C seasonal mean, with the season defined by all days with a daily mean temperature of at least 0.9 °C, and a minimum season length defined in this way of 94 days (about 3 months). The other assessment follows the idea of Walter and Medina (1969; see also Bendix and Rafiqpoor 2001) to "dig" for mean air temperatures by burying data loggers at 10 cm soil depth in 100% shaded sites. The soil-temperature data from around the world revealed a mean growing season air temperature of 6.4 °C at natural climatic tree lines. The average length of the growing season must not be less than 91 days. A longer growing season had no significant effect. The 0.3 K difference is plausible as a sign of global warming, since the data loggers measure recent temperatures, while WorldClim uses data for the period 1950–2000.

From the global climatology data of WorldClim (see Hijmans et al. 2005 and https://www.worldclim.org/data/worldclim21.html), **for the Ecoclimate Classification concept**, however, in ArcView 12 monthly maps of thermal growing season (**≥10 °C**) were derived. The summation of these maps resulted in a single **isothermomene map** (see Fig. 4.2) with 13 possible values (0 to 12 months), which were then divided into six ecologically relevant classes: 12 months = *megatherm*; 11–7 months = *macrotherm*; 6–5 months = *mesotherm*; 4–3 months = *microtherm*; 2–1 months = *oligotherm*; 0 months = *hekistotherm* (Greek: *hekisto* = least).

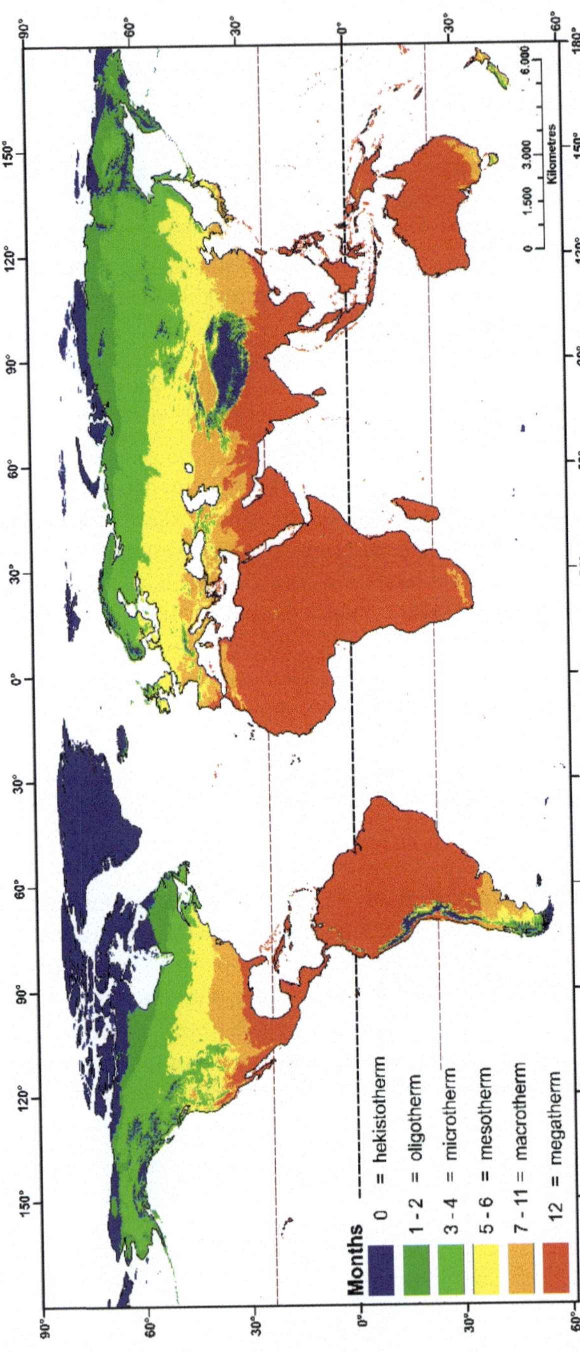

Fig. 4.2 Global patterns of the duration of the thermal growing season (= **Isothermomenes:** number of months with mean monthly temperature ≥10 °C)

4.1.1 *Spatial Pattern of the Thermal Dimension of Ecoclimate Classification (Isothermomenes)*

Isothermomenes give the potential length of the growing season, determined by the number of growing months ≥10 °C (see Sect. 4.1), which increases steadily towards the equator, parallel to the latitudinal increase in solar radiation during the year (Figs. 4.2 and 4.3a).

The most striking aspect in the **isothermomenes map** (see Fig. 4.2) is the dominance of the red colour (12 vegetation months) almost south of the Tropic of Cancer. Mean values per latitude band therefore tend to be lower in the Northern Hemisphere than at the same latitude in the Southern Hemisphere, due to the smaller landmass and higher climatic oceanity south of the equator (see Fig. 4.3a). Apart from breaks

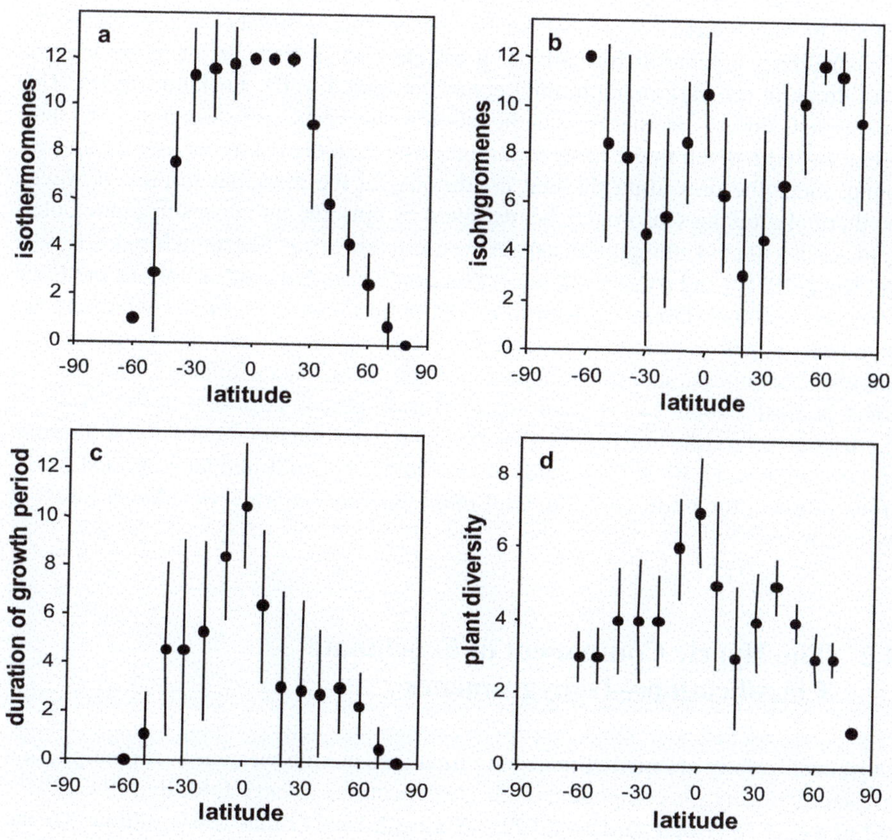

Fig. 4.3 Latitudinal variation of isothermomenes (**a**), isohygromenes (**b**), number of optimal hygrothermal growing months (**c**), and species richness of vascular plants (**d**); calculated after Barthlott et al. 2005) (−90 = S; 90 = N). Mean values and standard deviation per 10° latitudinal band are shown, except for species richness where the median of the ten diversity zones of the map by Barthlott et al. (2005) is shown

in mountainous regions, isothermal phenomena are largely latitudinal in nature. Furthermore, breaks in the latitudinal pattern are evident in regions characterised by oceanic climates (e.g., western Central Europe) or highly continental climates (e.g., eastern Asia, western North America). While the former cause poleward shifts of the isothermomenes, the latter cause equatorward shifts (see Fig. 4.2).

With the exception of the high mountain ranges, the megatherm climate (12 thermal vegetation months) covers most of the tropics and extends poleward into the adjacent subtropics, mainly in the oceanic regions on the eastern sides of the continents and on subtropical islands. Subtropical lowlands are mainly characterised by macrotherm climates. Temperate and boreal zones have mostly meso-, micro- and oligotherm climates. Hekistothermal regions are found poleward from the Arctic Circle and in high mountain regions in all climate zones (e.g., Rocky Mountains, Andes, Tibetan Plateau, Hindu Kush-Karakoram-Himalaya system).

While a greater variety of thermal gradations is found in the outer-tropics, the tropics are characterised by almost year-round thermally suitable conditions for the corresponding vegetation formations in all elevational belts, even if the thermal level drops in the various altitudinal zones compared to the lowland tropics. This means that, due to 'isothermy' in the tropics, the vegetation formations of individual altitudinal belt in the tropical mountains, i.e. in the Páramos, fructify all year round under the prevailing thermal conditions. In the marginal tropics, however, the thermal growing season may be restricted by episodic influences from the outer-tropics, such as cold air incursions from the north [known as ***Nortes*** in Mexico (Lauer and Klaus 1975)] and south polar advection [known as, ***Surazos*** in Bolivia and Peru, ***Friagems*** in Brazil (Pezza and Ambrizzi 2005; Latrubesse et al. 2009)], which are not adequately captured by monthly means (in the ecological climate diagrams periodic and absolute minima and maxima are recorded so far data available). The thermal effect is most evident in the subdivision of mid-latitude climates. In the Northern Hemisphere in continental regions, for example, the isothermomenes of four thermal growing months (≥10 °C) separates the microthermal cold-temperate boreal coniferous forest in the north from the cool-temperate mesothermal deciduous forest in the south.

4.2 The Hygric Component of Ecoclimate Classification—Isohygromenes

Most ecologically oriented climate classifications use the mean annual precipitation to quantify the hygric component of the classification system (Holdridge 1947, 1967; Whittaker 1975; Woodward et al. 1995; Woodward and Lomas 2004), although mean annual precipitation is of limited use in interpreting the moisture available to vegetation (Lauer 1952; Lauer and Frankenberg 1981a; Leemans et al. 1996). For example, the climate stations of **Makassar** (= Ujung Pandang/Indonesia: 05° 04′ S/119° 50′ E; 2 m asl) and **Iquitos** (Peru: 03° 45′ S/73° 18′ W; 125 m asl) are both characterised by high mean annual temperature (26.3 °C and 26.5 °C, respectively) and receive approximately similar high amounts of precipitation (3203 mm/a and 2879 mm/a, respectively) (Fig. 4.4). According to Whittaker's and Holdridge's classifications, both sites would fall into the same ecoclimatic category. Due to the winter monsoonal rainfall regime in Makassar (Indonesia), more than 80% of the annual precipitation falls between September and March. Iquitos (Peru), on the other hand, receives high amounts of rainfall throughout the year, with no distinct dry season. These seasonal differences of precipitation have drastic consequences for the potential vegetation (Lauer 1952): Monsoonal moist savannahs in Makassar surroundings; tropical evergreen forest in Iquitos. Using ecological climatic diagrams of Walter/Lieth for both stations (see Fig. 4.4, bottom) it is also very easy to see the essential difference at a glance. The Berlin diagram illustrates as legend to the ecoclimate diagrams the used colours for the various parameters.

An alternative—and ecologically more reasonable—way of capturing water availability for plant life is the concept of **water budget**, i.e. the balance between precipitation and evapotranspiration. Comparing the ecoclimate diagrams of Lauer and Rafiqpoor (2002) (see Fig. 4.4, top) with the climate diagrams of Walter and Lieth (1960–67) (see Fig. 4.4, bottom), it is noticeable that if potential landscape evapotranspiration (pET) is taken into account as an ecophysiologically significant parameter for plant metabolic processes (Lauer and Rafiqpoor 2002), if the vegetation cover withers during the short dry season (e.g., in Makassar) and can hardly transpire, the level of precipitation deficit for plant life also remains low. This aspect is indirectly reflected in the climate diagrams of Walter/Lieth, too (see Fig. 4.4, bottom). But these ecological climate diagrams use a rather simple approach, just by assuming the temperature curve as an indication for evaporation; months where the temperature curve is below the monthly precipitation amount are defined as relatively humid (indicated in the diagram by blue hatching); months with the temperature curve above the monthly precipitation amount as relatively arid (indicated by red stippling). Month with more than 100 mm are called hyperhumid and are indicated with darker blue and a shortening of the ordinate scale by the factor 10. This enables the reader to capture the climatic situation of a station or of a map with many stations (see Figs. 4.10 and 4.11) rather rapidly.

Since the considerations on the classification of climates by Penck (1910), potential evapotranspiration has been considered an important variable in ecoclimatology, macroecology and vegetation science. Potential evapotranspiration represents the ability of the atmosphere to remove water from a land surface by evaporation (from soils and rocks) and transpiration (from plant surfaces), assuming an unrestricted water supply. It is therefore considered a good proxy for the biologically relevant environmental energy in a given landscape.

The potential evapotranspiration depends strongly on the vegetation and its evaporative properties. This is because climatic evapotranspiration in a landscape, i.e. land covered by vegetation, strongly modifies the potential evaporation of the free water surface (Lauer and Frankenberg 1981b; Lauer and Rafiqpoor 2002). The **potential landscape evapotranspiration** (pLV) thus differs from the **potential evapotranspiration** (PET) sensu Thornthwaite (1943, 1948) in that the pLV does not assume

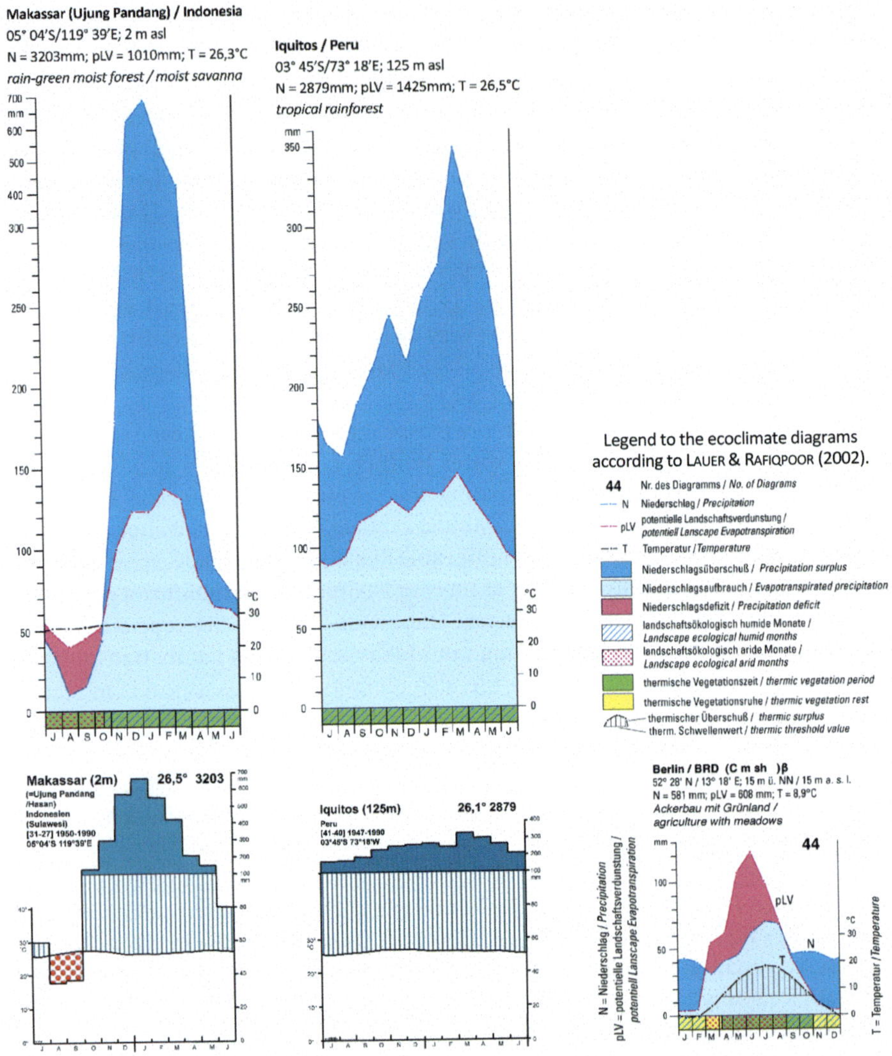

Fig. 4.4 Ecoclimate diagrams of Makassar and Iquitos according to the method of Lauer and Rafiqpoor (2002) (Fig. 4.4, top) and Waler/Lieth (1960–67) (Fig. 4.4, bottom) for two different tropical stations and temperate Berlin as examples to improve the informative value of a climate classification (details see also Figs. 4.10 and 4.11)

a uniformly dense vegetation cover (e.g., short cut lawn), but takes into account the real ecotope conditions of soil's **evaporation** and vegetation's **transpiration**. Furthermore, unlike Penman's (1948) concept, it does not assume that the transpiration conditions of the vegetation are always uniform, but takes into account the seasonal changes in the optimal transpiration of real and potential vegetation units in a landscape (Lauer and Frankenberg 1981b; Lauer and Rafiqpoor 2002). In addition, the Lauer and Frankenberg (1981b) and Lauer and Rafiqpoor (2002) methodologies require data on equivalent temperature, barometric pressure, saturation deficit, relative humidity … in order to calculate pLV. Unfortunately, these data are not yet available in digital form with sufficient accuracy on a global scale.

In this study, therefore, we used the PET data of FAO and defined hygrically suitable growing months as those with a positive **water balance** (N ≥ PET). Based on monthly data for **precipitation** (**N**: from WorldClim: Hijmans et al. 2005: see https://www.worldclim.org/data/worldclim21.html) and **potential evapotranspiration** (**PET**: from FAO dataset: Zomer & Trabucco 2022: see https://doi.org/https://doi.org/10.6084/m9.figshare.20005589), in GIS we calculated Earth's water balance (**WTB = N–PET**) maps for each month of the year. This approach combines both the amount and the timing of precipitation into a single ecologically relevant variable. Months with a positive water balance were considered as **humid** if the water balance in the corresponding month was **≥0**, otherwise the corresponding month was classified as **arid**, having a negative water balance. The addition of the 12 monthly maps resulted in an **isohygromene** map with 13 classes (0–12 months), i.e. areas with the same number of humid months. This map was reclassified according to vegetation types into six categories of landscape ecological relevance: 12–10 months = *hyperhumid*; 9–7 months = *euhumid*; 6–5 months = *subhumid*; 4–3 months = *semiarid*; 2–1 months = *euarid*; 0 months = *hyperarid* (see Fig. 4.5).

4.2.1 *Spatial Pattern of the Hygric dimension of Ecoclimate Classification (Isohygromenes)*

As mentioned above, Isohygromenes indicate the potential length of the growing season as a function of water availability. In contrast to isothermomenes, the latitudinal pattern of isohygromenes appears to be much more complex, especially in the tropics (Fig. 4.5). High mean values per latitude band are found in the inner-tropics as well as at higher latitudes (see Fig. 4.3b). The lowest mean values are found in the subtropics, where the variability is also accentuated.

Furthermore, a slightly asymmetric latitudinal decrease in mean isohygromene values can be observed, with the subtropics of the northern hemisphere being slightly drier than their southern counterparts (see Fig. 4.3b), paralleling the hemispheric asymmetry of thermal conditions.

While the entire outer-tropics (subtropical, temperate and polar zones) are mainly thermally differentiated, the tropics are mainly hygrically differentiated. Although sufficient energy may be available, a negative water balance during the dry season in most parts of the outer-tropics can lead to an interruption of growth in these otherwise thermally optimal regions. All hygric climate types are found in the tropical regions (see Fig. 4.8, zone **A** and see Fig. 4.9 as legend to Fig. 4.8) of all continents. Hyperhumid conditions occur, for example, in the Amazon basin, the Congo basin, coastal West Africa and much of the Indo-Malayan archipelago, but also in some

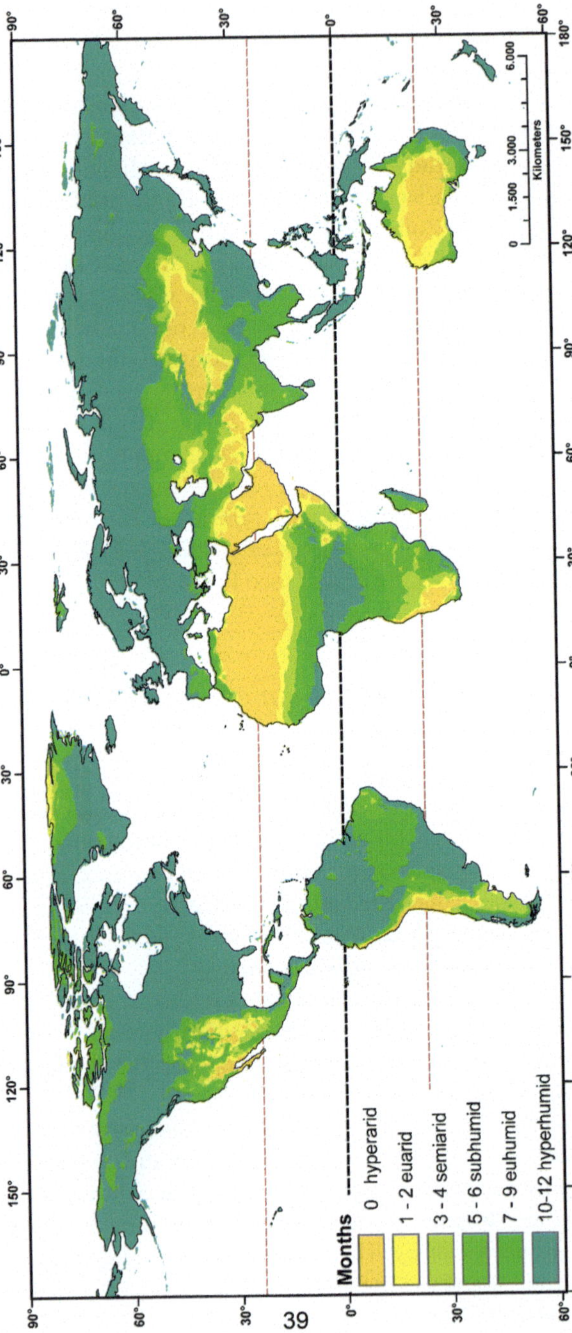

Fig. 4.5 Global patterns of the duration of the hygric growing season (= **Isohygromenes**: number of months with positive water balance)

oceanic regions of the outer tropics and in luv-sides of mountains. On the other hand, hyperarid tropical and subtropical regions are limited to the coastal (Atacama, Namib) and inland deserts (Sahara, Horn of Africa, Arabian Peninsula), which are differentiated into tropical and subtropical dry climate types by the corresponding solar climatic boundaries and annual temperature oscillations (see Fig. 4.8).

In the subtropics (see Fig. 4.8, zone **B**), the hygric component has a much greater influence than the thermal, except in the more east-west oriented high mountains where the thermal component is dominant. In the temperate (see Fig. 4.8, zone **C**) and polar (see Fig. 4.8, zones **D** and **E**) zones, euhumid growing conditions dominate due to sufficient water supply combined with limited energy input and consequently lower evapotranspiration, although areas with a dry climate also occur here. Again, highly continental regions and leeward sides of mountain ranges (e.g., western North America) show opposite trends. The eastern sides of the continents in Asia, North America and south-eastern Australia are hyperhumid, while hyperarid conditions are confined to the interior of the continents, such as in high Asia between the Tarim Basin and the Gobi Desert.

4.3 The Hygrothermal Dimension (Duration of Optimal Growing Season) of Ecoclimate Classification

The combination of the monthly values of isothermomene and isohygromene data in GIS provides a map of the duration of the optimal growing season with coinciding thermally and hygrically optimal conditions for plant life (Fig. 4.6).

On a global scale, the number of hygrothermally optimal growing months decreases with latitude, with the lowest variability at higher latitudes and the greatest variability in the subtropics (see Fig. 4.3c).

The humid tropics show all year-round optimal conditions with up to 12 optimal growing months. On the other hand, deserts and semi-deserts, as well as high arctic regions and high arctic and mountain deserts, have unfavourable ecoclimatic conditions for plant life throughout the year. Between these extremes, there are many strictly latitudinal transitions, such as the transition from the Sahara to the Congo Basin in Africa or in eastern North America. On the other hand, longitudinal arrangements are found along longitudinally arranged mountain ranges (e.g. Rocky Mountains, southern Andes). Mediterranean regions have in general an intermediate duration of the optimal hygrothermal growing season, but are highly differentiated at regional scales in mountainous regions.

In addition, some parts of the western and eastern sides of continents in the tropics, which are influenced by monsoonal effects, have particularly suitable conditions (e.g., Chocó, Cameroon-Guinea, Western Ghats, Himalayas, south-east China and Indo-Malayan Archipelago). The same applies to the eastern flanks of continents in the inner and outer-tropics, and in some cases in the subtropics, which are under the constant influence of the trade winds (luv effect) and are therefore generally

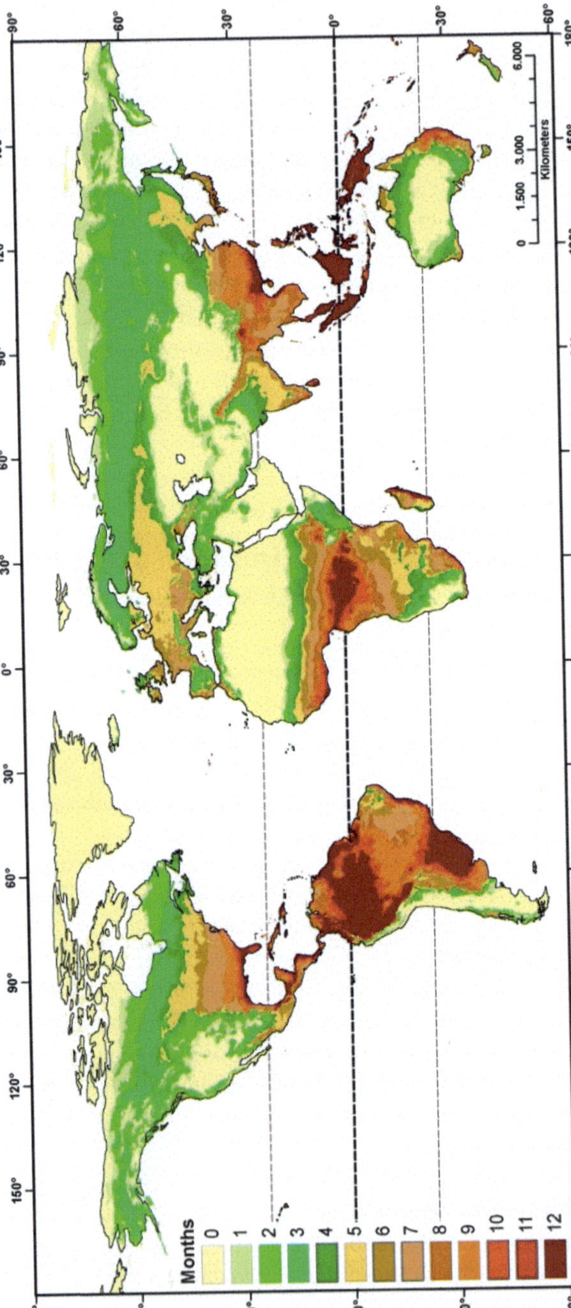

Fig. 4.6 Global patterns of the hygrothermal growing season (number of months which have both optimal thermal and hygric growth conditions)

wetter than the western flanks at the same latitudes: Eastern flanks of the tropical Andes, Central American isthmus and islands, Mata Atlantica of eastern Brazil, eastern Madagascar, eastern Australia, northern New Zealand, Uruguay-Argentinean coastal regions, south-eastern China. This effect is further enhanced by the influence of mountain ranges due to the damming effect.

4.4 The Variation of Day-Length (DLV)

In the ecoclimate classification map (see Fig. 4.8) presented here, the solar-climatic boundaries of the main climate zones (**A** = tropical, **B** = subtropical, **C** = temperate, **D** = boreal, **E** = polar) are shown according to the latitudinal annual day length variation, following the concept of Lauer et al. (1996) and Lauer and Rafiqpoor (2002). Solar parameters in the form of quantitatively determinable thresholds of the annual daylight variation (DLV) are used to differentiate the climate zones. The DLV can be calculated for both hemispheres from the difference between the length of daylight at the summer and winter solstices. Sunrise and sunset, which determine the length of daylight, can be derived from the nomogram (Fig. 4.7) from the equator to the poles for selected dates. The code numbers in Fig. 4.7 refer to the dates of the equinoxes (**II**) and the winter and summer solstices (**I, III**). The solar-climatic boundaries between the major climate zones (**23.5°** = tropical/subtropical with a DLV = 3 h; **45°** = subtropical/temperate with a DLV = 7 h; **66.6°** = boreal/polar with a DLV = 24 h) run parallel to the latitudes according to the astronomical law (Lauer et al. 1996).

If, when defining the five major climate zones (**A, B, C, D, E**) in the context of an **Ecoclimate Map** (see Fig. 4.8), geographically relevant parameters such as land-sea distribution, ocean currents, relief (with its different altitudes and exposures), and energy transport in the atmosphere are taken into account, these boundary belts on the map deviate to a "tolerable" degree from their ideal line (Louis 1958) and basically correlate widely also with the picture of tropospheric atmospheric circulation (Lauer 1991; Lauer and Rafiqpoor 2002; Rafiqpoor 2024).

Although the solar boundaries are strictly latitudinal, the corresponding climatic boundaries indicate embayments, especially on the east and west sides of the continents, mainly because of the above-mentioned phenomena. Thus, the boundaries of the climate zones differentiate the climates with the same hygrothermal conditions (with the same legend colour: see Figs. 4.8 and 4.9) into two different climate types with respect to the climate zones [see, for example, the megatherm-hyperhumid tropic and subtropic climates on either side of the Tropic of Capricorn on the eastern side of South America or at the tropics/subtropics boundary in Florida]. Taking into account the lines between the large-scale climate zones in a climate classification map therefore helps to avoid assigning climates with similar hygrothermal properties (**"Homoclimates"**) to an inappropriate climate zone, which is done in the classification of Köppen (1900), as well as in the later digital classifications of Kottek et al. (2006), Peel et al. (2004) and Stern et al. (2000).

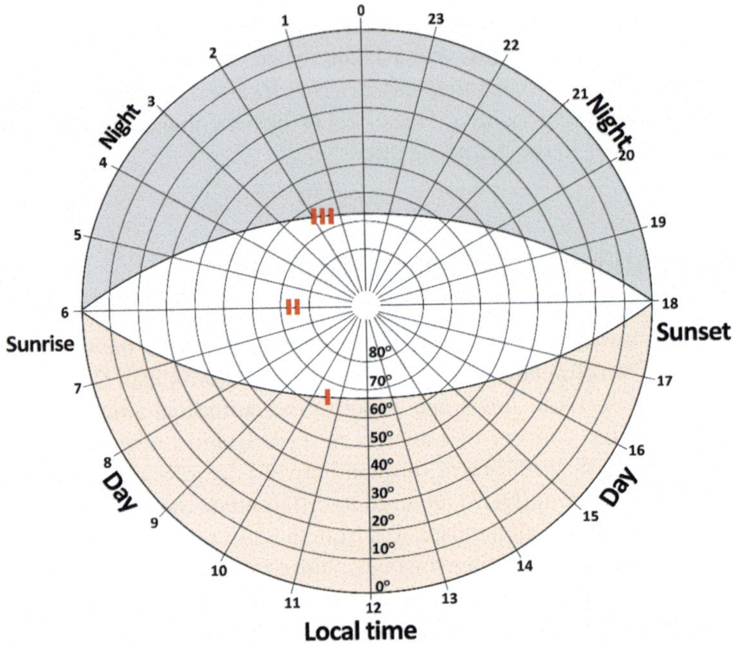

Northern hemisphere	Date	Southern hemisphere
I	December 21th, Winter solstice	III
II	March 20th and September 22nd: Equinoxial times	II
III	June 21th: Summer solstice	I

Fig. 4.7 The duration of the theoretical daily sunlight in dependence on the geographical latitude (varied from Junghans 1969). The given data for the astronomical events may vary by +/- one day from year to year

4.5 The Ecoclimate Classification—A Novel Synthetic Approach

Cross-tabulation of the unclassified maps of isohygromenes and isothermomenes yields 13 × 13 = 169 theoretical combinations of climate types. For the sake of clarity, the 169 combinations of isothermomenes and isohygromenes were further reclassified into 36 ecoclimate types (Fig. 4.8) according to Lauer and Rafiqpoor (2002), using the raster calculator of ArcView. The result is a world map of the ecoclimate classification (see Fig. 4.8) with quantitatively defined climate types, which, in contrast to most existing classifications, also assigns quantitatively defined climate types to the world's high mountain regions.

The map of the ecoclimate classification is rather self-explanatory due to the clarity of the three main delimitation criteria (isothermomenes, isohygromenes, day length variation). However, a general view shows that the main colours in the legend (see Figs. 4.8 and 4.9) express the latitudinal pattern of the main climatic zones delimited by the radiation-climate defined **DLV**, although the same hygrothermal

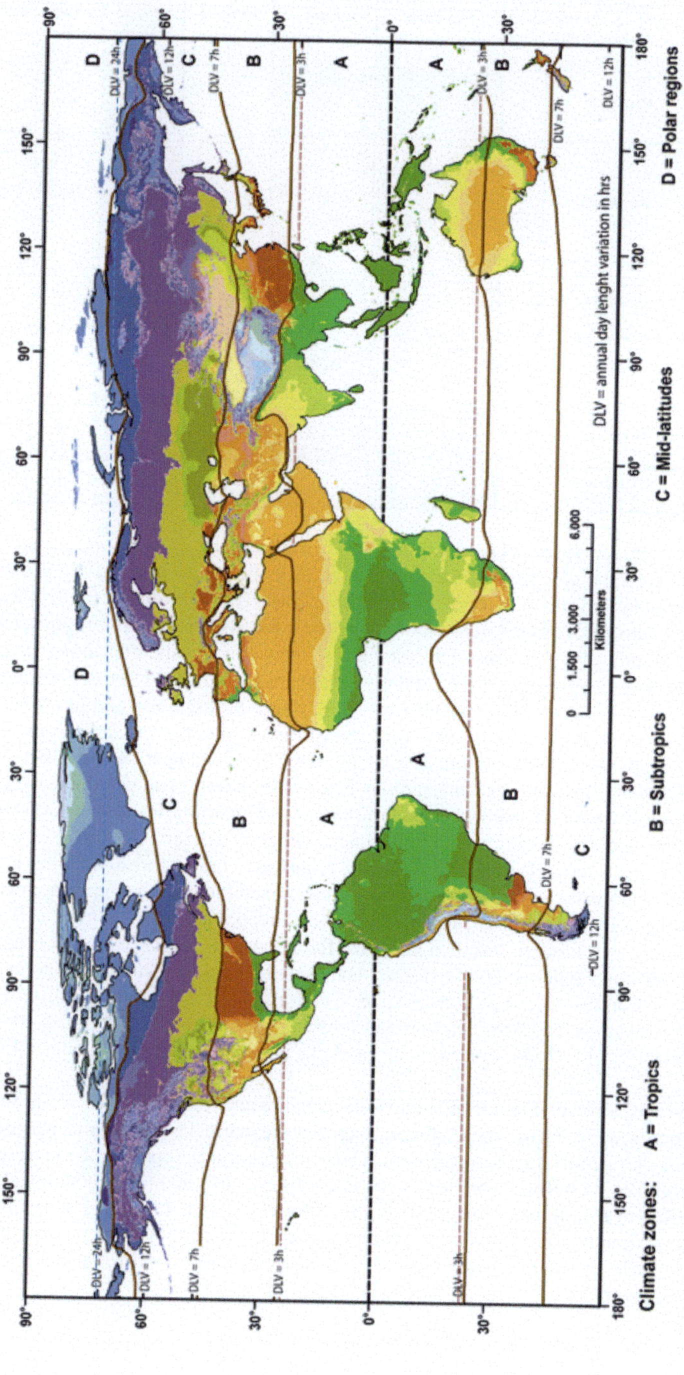

Fig. 4.8 Ecoclimatic classification of the Earth (for detailed legend see Fig. 4.9)

CLIMATIC ZONES	Heat budget: number of months allowing thermal plant growth		Water budget: number of humid months allowing hygric plant growth					
			hyperhumid	euhumid	subhumid	semiarid	euarid	hyperarid
			12 11 10	9 8 7	6 5	4 3	2 1	0
A TROPIC	Megatherm (in all climatic belts)	12	01	02	03	04	05	06
23,5° N and S B SUBTROPIC	Macrotherm	11 10 9 8 7	07	08	09	10	11	12
45° N and S C TEMPERATE	Mesotherm	6 5	13	14	15	16	17	18
57° N and S D BOREAL	Microtherm	4 3	19	20	21	22	23	24
66,6° N and S E POLAR	Oligotherm	2 1	25	26	27	28	29	30
glaciated areas in A, B, C, D, E	Hekistotherm (in all climate zones)	0	31	32	33	34	35	36

Fig. 4.9 Legend of the ecoclimate classification of the World based on isothermomenes and isohygromenes; colours correspond to map (see Fig. 4.8). Figures in the colour fields of Fig. 4.9 indicate the natural vegetation formations in the corresponding ecoclimate types as follows: **01**: Tropical evergreen rain forests, subtropical moist forests, moist *Araucaria* forests in S-Brazil-Uruguay, Pantanal (Paraná basin); **02**: semi-evergreen moist forests, moist monsoon forests, and moist savannahs; **03**: Dry forests, deciduous monsoon forests, and dry savannahs (Miombo, Mopane, Caatinga, Gran Chaco, *Eucalyptus* forests in Australia); **04**: Dry deciduous thorn shrubs, thorn- and succulent savannahs; **05**: Semi-deserts; **06**: Deserts; **07**: Subtropical moist forests, moist grasslands (Pampa, Plains in N-America), warm-temperate mixed forests (Colchis, the Crimea, Hungarian basin, Biscay bay); **08**: Subtropical sclerophyllous and/or lauriphyllous forests; **09**: Subtropical deciduous woodlands and steppes; **010**: Short grass and scrub steppes (Prairie, Central Australia, Rajasthan, E-Patagonia); **11**: Subtropical winter cold semi-deserts; **12**: Subtropical winter cold deserts (Tarim basin, Salt Lake City); **13**: Summer green broad leafed mixed forests, temperate conifer forests (Pacific forests in N-America, Valdivian forests , S-Chile); **14**: Summer green mixed forests and temperate grasslands; **15**: Winter cold steppes of mid-latitudes; **16**: Winter cold dry steppes and grass formations(E-China, W-Patagonia); **17**: Temperate semi-deserts (Gobi); **18**: Temperate deserts; **19**: Boreal conifer forests (Taiga); **20–23**: Transition types of conifer forests and grass steppes (N-China, Mongolia, Tibet); **24**: Cold deserts of Central Asia high plateaus; **25**: Tundra; **26–30**: Different transition types of vegetation patches, swamps, and frost debris in the subpolar regions and high altitudes all over the world; **31**: Frost debris belt; **32–36**: Polar ice caps and glaciated areas in high altitudes all over the world

The climate diagrams corresponding to the individual ecoclimate types in Fig. 4.9 are arranged on the following pages (see Fig. 4.10). These climate diagrams are marked with the same number as the climate type in Fig. 4.9 (from 01 to 36 as follows: **ET-01** = Ecoclimate Type 01), the corresponding zonobiomes are marked with Roman numerals. In these figures, 2 to 3 additional ecological climate diagrams **according to zonobiome concept of Walter/Lieth** (see Sect. 4.6), as far as available, are shown for the corresponding ecoclimate type from other regions of the world to demonstrate the distribution of the corresponding climate type worldwide (with indication of relevant zonobiome (ZB), or zonoecotone (ZE) ecoclimate types can also occur in other climatic belts. Furthermore, in the tropics (zone **A**), the latitudinal sequences of ecoclimates are found in regions where the hygric component is also latitudinal (e.g., N-S transect from the Sahara to the Congo Basin). In the subtropical belt (zone **B**) there is a considerable diversity of ecoclimate types. The main drivers of this great heterogeneity are mountain ranges and the effects of oceanity or continentality. The temperate, boreal and polar regions, on the other hand, are characterised by largely latitudinal sequences of mesotherm to hekistotherm climates under mostly euhumid to hyperhumid conditions, but also here exist cold-euarid regions.

However, the ecoclimate classification model presented here has the potential to infer changes in vegetation cover from changes in climate and vice versa. The modelling of the shift in climate zones as an expression of global climate change (see Part II) will be illustrated in section 2.5 using the IPCC 'SRES **A2**' scenario with a temperature increase of 3.4 K for the 21st century (2050).

4.6 The Zonobiome Concept and Ecological Climate Diagrams

Another concept in climate classification was the attempt to differentiate the earth into ecological Regions. Walter (1960, Walter 1979a) was the first to divide the earth into ecological zones (so-called zonobiomes and ecotones or zonoecotones). After more than 60 years of ecological research, he concluded that the division of the Earth into vegetation zones alone is not purposeful, because vegetation is only one, albeit particularly important, living component of ecological systems; it should not be treated in isolation from the other components, of which **climate** is the most important and independent (Walter and Breckle 1991). Thus, Walter developed a **climate diagram system** to characterize **'Homoclimates'** and divided the Earth into ecological zones (zonobiomes).

The climate diagram system (see Walter and Lieth 1960–67) is characterised by the use of monthly temperature data as an approximation for the evaporative demand. This approach goes back to Gaussen (1954), who had tested it for the Mediterranean region. He found that the ratio of 10 °C≙20 mm of rain (per month) corresponds quite well to the evaporative demand (see above PET and temperature correlation: Fig. 4.1). Thus, it gives a rough idea of the water balance, i.e. the distinction between relatively arid and relatively humid seasons. This could be demonstrated for many other climatic regions, but in steppe areas or in very windy regions (i.e., Patagonia) the ratio 10 °C≙30 mm gives a better idea of dry periods that do not reach the severity of a drought or a real dry season.

Climatecharts.net is an interesting source for climatic data from all over the world at the internet. They provide data from two different kind of sources, mostly interpolated and simulated climate data for raster cells on the map and then also some real measured weather data from weather stations around the world. Temperature and precipitation values shown in the climate diagrams on the website are based on interpolated data and therefore might differ significantly from real local conditions, which is a disadvantage. Furthermore, the diagrams which they claim to be Walther-Lieth climate charts have only an eleven-month abscissa per year and the proportion between abscissa and ordinate scale is also different from the original Walter-Lieth-Harnickell-diagrams (1960) resp. those in use here. Anyhow it is possible to play in many ways to see completeness of data for varied chosen periods or the statistic distribution of temperature and precipitation values with box-plots etc. (Zepner et al. 2020).

Ecological climate diagrams and corresponding maps are now in use throughout the world as means of quickly and clearly characterising the climate of a particular region or showing typical climatic gradients. Climate diagrams now in use are described below (see Fig. 4.10 ff). Most data were taken from http://www.worldclimate.com/.

The use of ecological climatic diagrams or delimitation of ecological zones, as shown by Walter and Lieth (1960–67) resulted in nine zonobiomes with respective zonoecotones as transition zones in between. Table 4.1 provides an overview of the nine zonobiomes (ZB) (see Fig. 4.10). In this illustration, the ecoclimate types (ET) are represented by an ecoclimate diagram (where available) according to the method of Lauer and Rafiqpoor (2002). These ecoclimate diagrams are accompanied by 2–3 climate diagrams according to Walter and Lieth (1960–67) with improved layout for the identification of zonobiomes. Each zonobiome has its set of corresponding vegetation and soil types (Breckle and Rafiqpoor 2022). This is a simpler approach as that of Lauer and Rafiqpoor (2002), but the similarity to the hygrothermal growing season map shown above is obvious. Thus, both approaches are complementary; there is clear congruence in the zonation of the two systems (see Table 4.1).

Table 4.1 Overview of the 9 zonobiomes, as shown by Walter and Lieth (1960–67)

ZB I =	Megatherm, hyperhumid climates of the inner tropics, daily temperature amplitude larger than annual temperature fluctuation
ZB II =	Megatherm, euhumid climate of the outer tropics, rainy season in summer, daily temperature amplitude similar than annual temperature fluctuation
ZB III =	Megatherm-macrotherm, arid desert climates of the marginal tropics and subtropics, annual temperature amplitude larger than daily temperature fluctuation (ZB III-IX)
ZB IV =	Macrotherm, semiarid, semi-humid, euhumid climates of the Mediterranean winter rain areas
ZBV =	Macrotherm, hyperhumid-euhumid subtropical climates (mostly) of the eastern sides of the continents, cool winter period
ZB VI =	Mesotherm, hyperhumid-semihumid climates of the cool mid-latitudes, cool to cold winters
ZB VII =	Mesotherm, semiarid and euarid climates of the cool mid-latitudes, cold winter period
ZB VIII =	Microtherm, hyperhumid-semihumid or semiarid climates of the cold mid-latitudes, long cold winter period
ZB IX =	Oligotherm, hyperhumid-semihumid climates of the cold subpolar and polar regions, frosty months most of the year

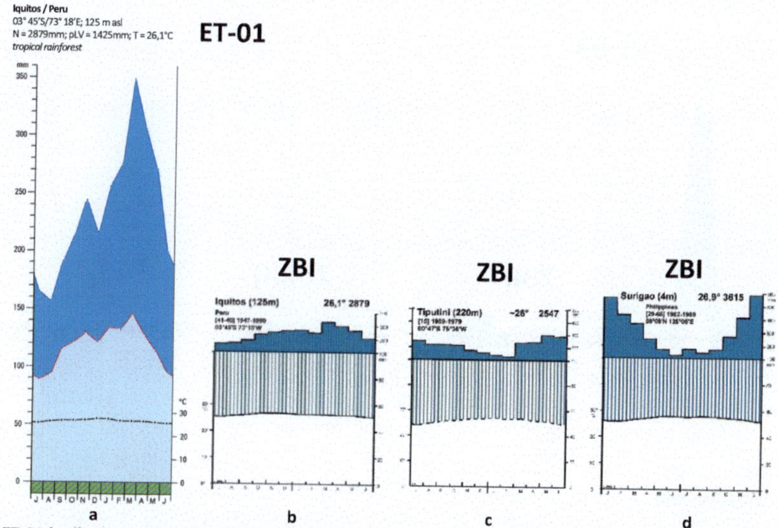

ET-01 (a-d): Tropic: Megatherm/Hyperhumid – Tropical evergreen rain forests, subtropical moist forests, moist *Araucaria* forests in S-Brazil-Uruguay, Pantanal (Paraná basin): Iquitos/Peru {ZBI, Fig. a, b}; Tiputini/Ecuador {ZBI, Fig. c}; Surigao/Philippines {ZBI, Fig. d}

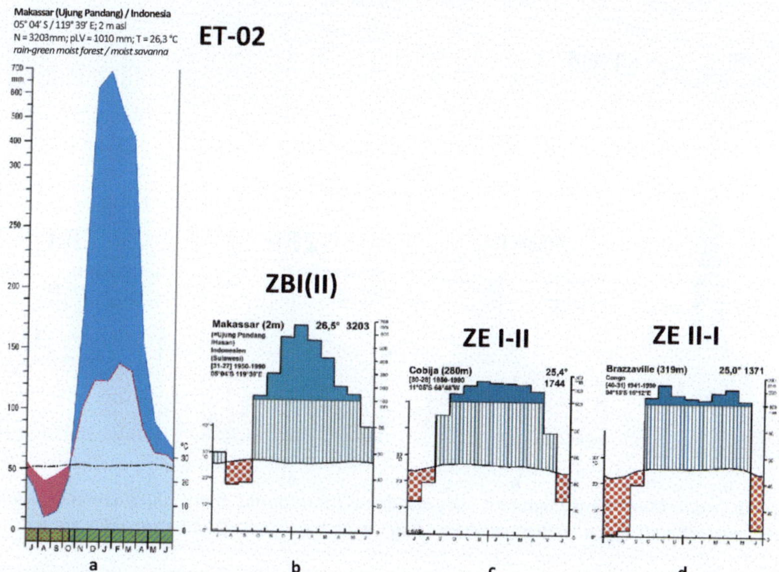

ET-02 (a-d): Tropic: Megatherm/Euhumid – semi-evergreen moist forests, moist monsoon forests, and moist savannahs: Makassar/Indonesia {ZBI(II), Fig. a, b}; Cobija/Bolivia {ZEI-II, Fig. c}; Brazzaville/Kongo {ZEII-I, Fig. d}

Fig. 4.10 Climate diagram-pairs according to the methods of Lauer/Rafiqpoor (always left) and Walter/Lieth, Breckle (always right) for the ecoclimate types, shown in Figs. 4.10 and 4.9 (= legend to Fig. 4.8). Some additional climate diagrams are shown for other parts of the world with same climate type. **ET** = Ecoclimate Type; **rIII**: Rain scarce as in ZBIII; **tI**: Temperature as in ZBI; **ZB** = Zonobiome; **ZE** = Zonoecotone. As a legend to ecoclimate diagrams (Lauer and Rafiqpoor) see Fig. 4.4

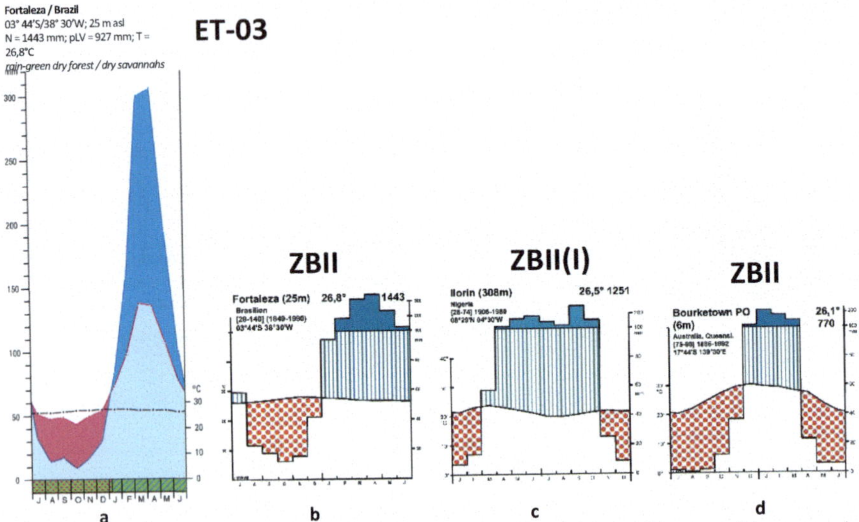

ET-03 (a-d): Tropic: Megatherm/Subhumid – Dry forests, deciduous monsoon forests, and dry savannahs (Miombo, Mopane, Caatinga, Gran Chaco, *Eucalyptus* forests in Australia): Fortaleza/Brazil {ZBII, Fig. a, b}; Ilorin/Nigeria {ZBII(I), Fig. c}; Bourketown/Australia {ZBII, Fig. d}

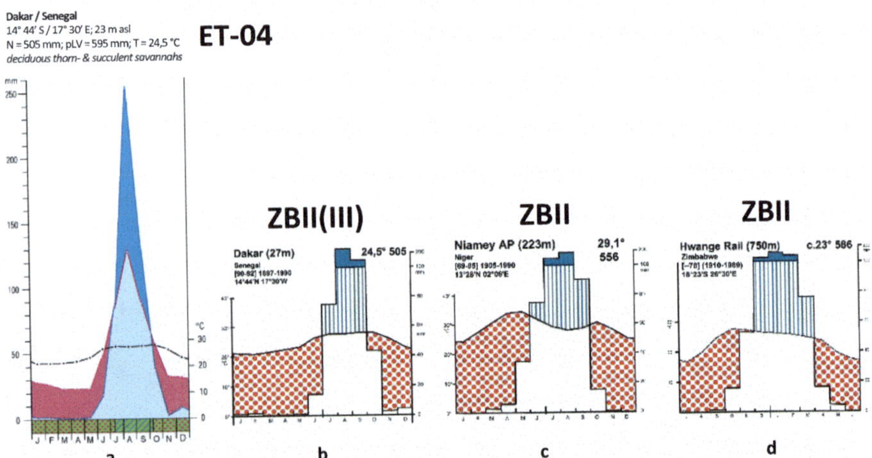

ET-04 (a-d): Tropic: Megatherm/Semiarid – Dry deciduous thorn shrubs, thorn- and succulent savannahs: Dakar/Senegal {ZBII(III), Fig. a,b}; Niamey/Niger {ZBII, Fig. c}; Hwange Rail/Zimbabwe {ZBII, Fig. d}

Fig. 4.10 (continued)

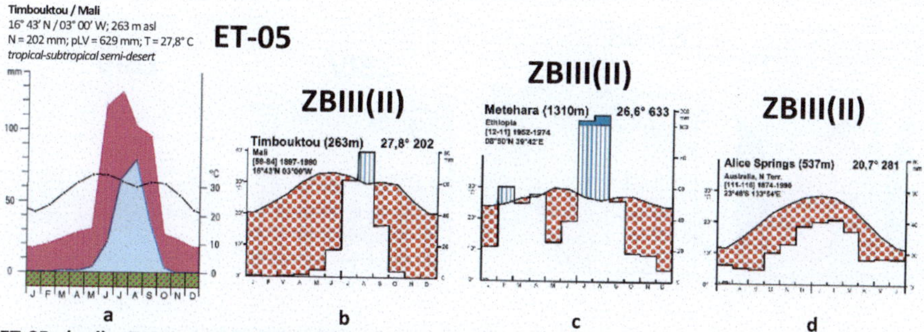

ET-05 (a-d): Tropic: Megatherm/Euarid – Semi-deserts: Tombouktou/Mali {ZBIII(II), Fig. a, b}; Metehara/Ethiopia {ZÖII-III, Fig. c}; Alice Springs/Australia {ZBIII(II), Fig. d}

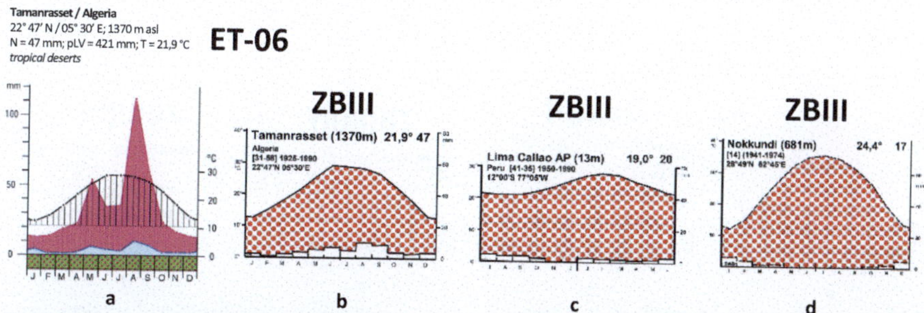

ET-06 (a-d): Tropic: Megatherm/Hyperarid – Deserts: Tamanrasset/Algeria {ZBIII, Fig. a, b}, Lima/Peru {ZBIII(tl), Fig. c}; Nokkundi/Iran {ZBIII, Fig. d}

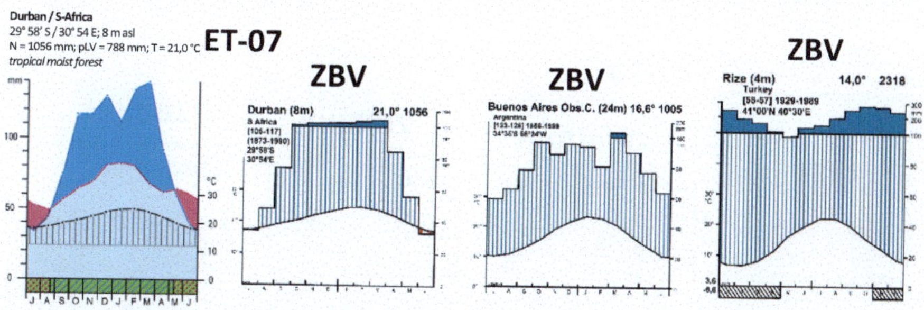

ET-07 (a-d): Subtropic: Macrotherm/Hyperhumid – Subtropical moist forests, moist grasslands (Pampa, Plains in N-America), warm-temperate mixed forests (Colchis, Biscay Bay): Durban/S-Africa {ZBV, Fig. a, b}; Buenos Aires/Argentina {ZBV, Fig. c}; Rize/Turkey {ZBV, Fig. d}

Fig. 4.10 (continued)

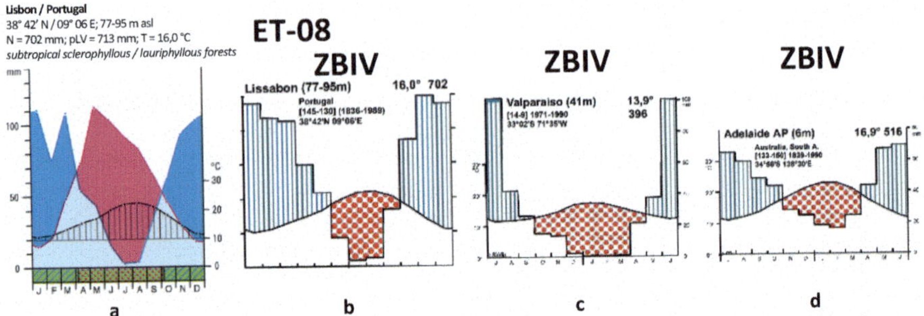

ET-08 (a-d): Subtropic: Macrotherm/Euhumid – Subtropical sclerophyllous and/or lauriphyllous forests; Medit., California, Middle Chile, S-Australia, S-Africa: Lisbon/Portugal {ZBIV, Fig. a, b}; Valparaiso/Chile {ZBIV, Fig. c}, Adelaide/Australia {ZBIV, Fig. d}

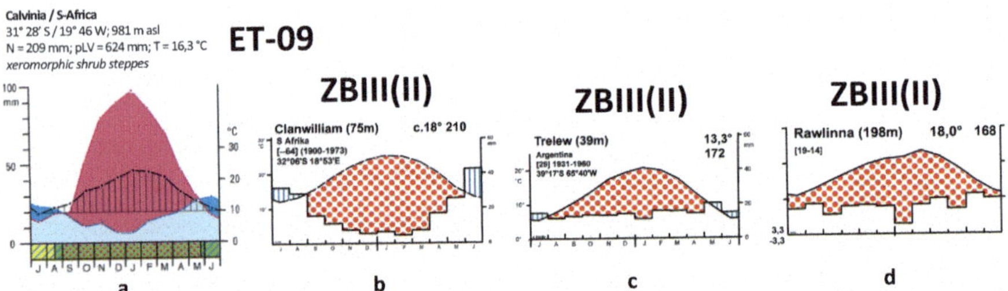

ET-09 (a-d): Subtropic: Macrotherm/Subhumid – Subtropical deciduous woodlands and steppes: Calvinia/S-Africa {ZBIII, Fig. a}; Clanwilliam/S Africa {ZBIII, Fig. b}; Trelew/Argentina {ZBIII(II), Fig. c}; Rawlinna/Australia {ZBIII(II), Fig. d}

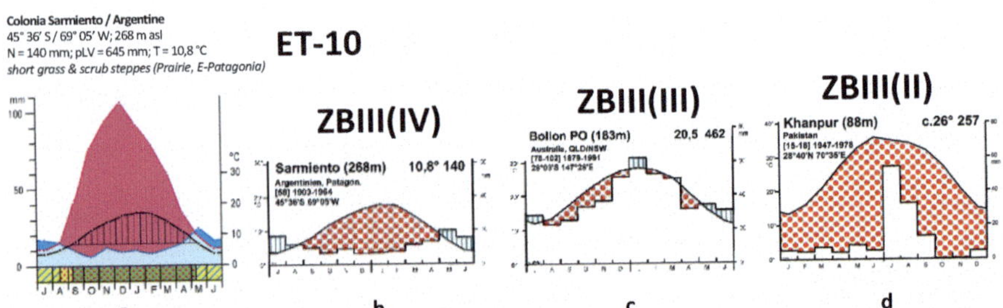

ET-10 (a-d): Subtropic: Macrotherm/Semiarid – Short grass and scrub steppes (Prairie, Central Australia, Rajasthan, E-Patagonia): Colonia Sarmiento/Argentina {ZBIII(IV), Fig. a, b}; Bollon/Australia {ZBII(III), Fig. c}; Khanpur/Pakistan {ZBIII(II), Fig. d}

Fig. 4.10 (continued)

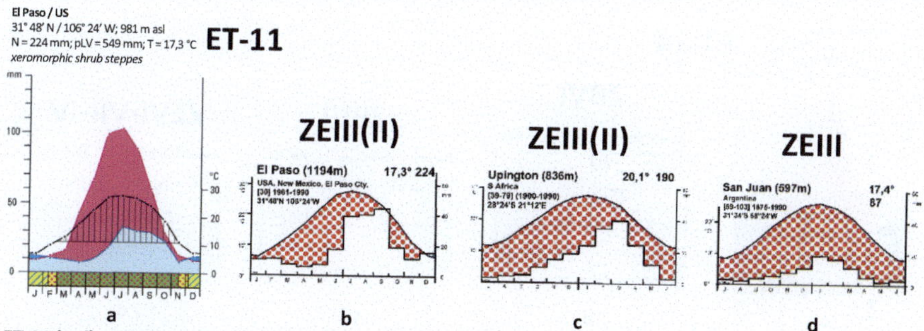

ET-11 (a-d): Subtropic: Macrotherm/Euarid – Subtropical winter cold semi-deserts: El Paso/USA {ZEIII-II, Fig. a, b}; Fig. a, b}; Upington/S Africa {ZEIII(II), Fig. c}; San Juan/Argentina {ZEIII, Fig. d}

ET-12

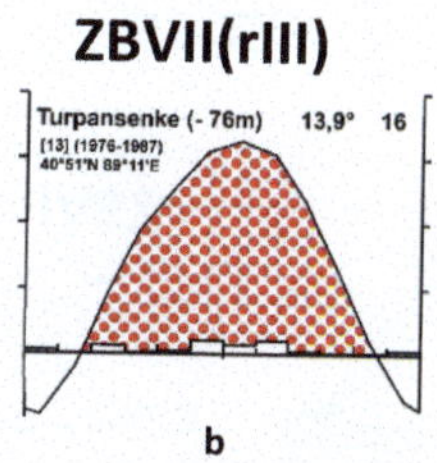

ET-12 (b): Subtropic: Macrotherm/ Hyperarid – Subtropical winter cold deserts (Tarim basin): Turpan Basin/Sinkiang {ZBVII(rIII), Fig. b}

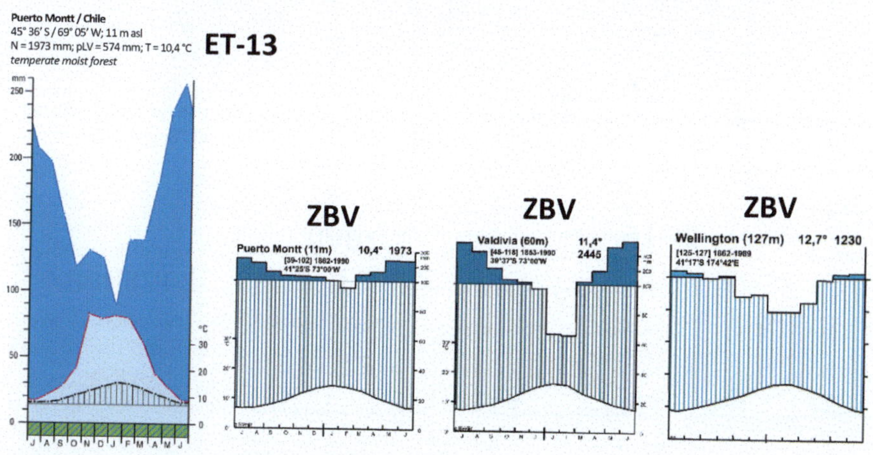

ET-13 (a-d): Temperate: Mesotherm/Hyperhumid – Summer green broad leafed mixed forests, temperate conifer forests (Pacific forests in N-America, Valdivian forests, S-Chile): Puerto Montt/Chile {ZBV, Fig. a, b}; Valdivia/Chile {ZBV, Fig. c}; Wellington/New Zealand {ZBV, Fig. d}

Fig. 4.10 (continued)

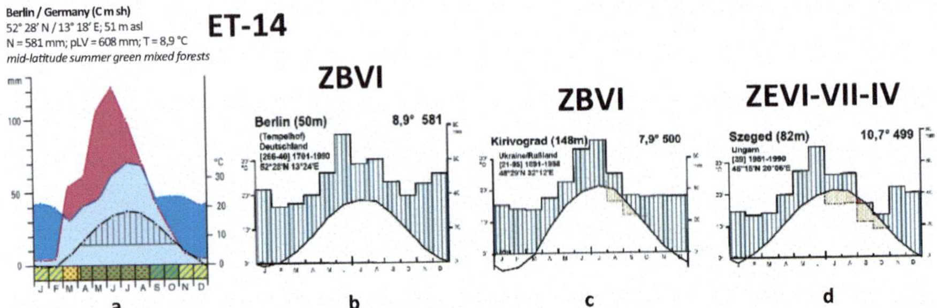

ET-14 (a-d): **Temperate: Mesotherm-Euhumid** – Summer green mixed forests and temperate grasslands: Berlin/Germany {ZBVI, Fig. a, b}; Kirowograd/Ukraine {ZBVI, Fig. c}; Szeged/Hungary {ZEVI-VII-IV, Fig. d}

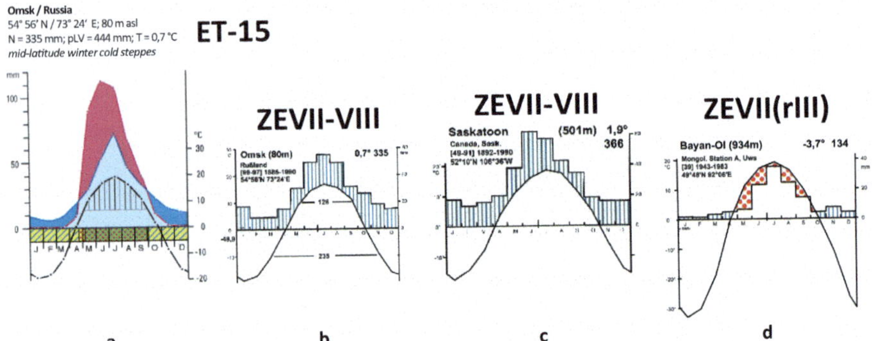

ET-15 (a-d): Temperate: Mesotherm/Subhumid – Winter cold steppes and dry conifer forests of mid-latitudes: Omsk/Russia {ZEVII-VIII; Fig. a, b}; Saskatoon/Canada {ZEVII-VIII, Fig. c}; Bayan-Ul/Mongolia {ZEVII(rIII), Fig. d}

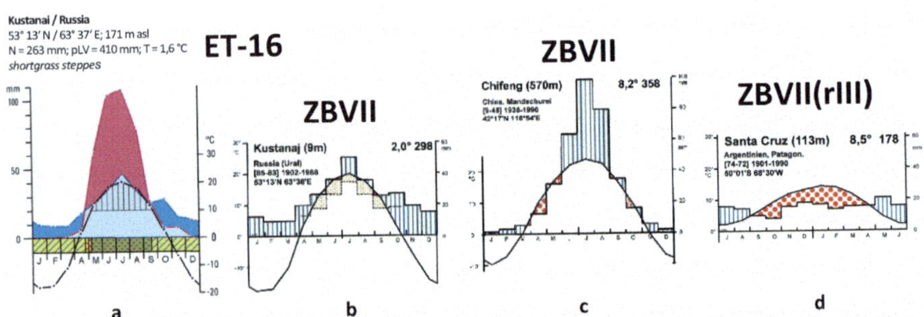

ET-16 (a-d): Temperate: Mesotherm/Semiarid -- Winter cold dry steppes and grass formations (E-China, W-Patagonia): Kustanai/Russia {ZBVII, Fig. a, b}; Chifeng/E-China {ZBVII, Fig. c}; Santa Cruz/Argentina {ZBVII(rIII), Fig. d}

Fig. 4.10 (continued)

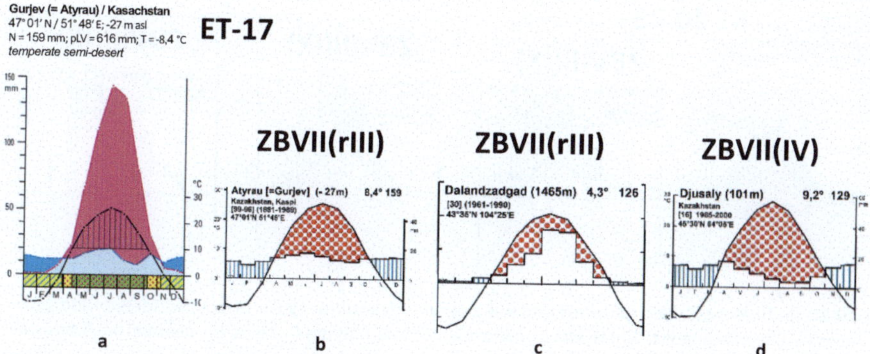

ET-17 (a-d): Temperate: Mesotherm/Euarid -- Temperate semi-deserts: Gurjev (= Atyrau)/Kazakhstan {ZBVII(rIII), Fig. a, b}; Dalangzadgad/China {ZBVII(rIII), Fig. c}; Djusaly/Kazakhstan {ZBVII(rIII), Fig. d}

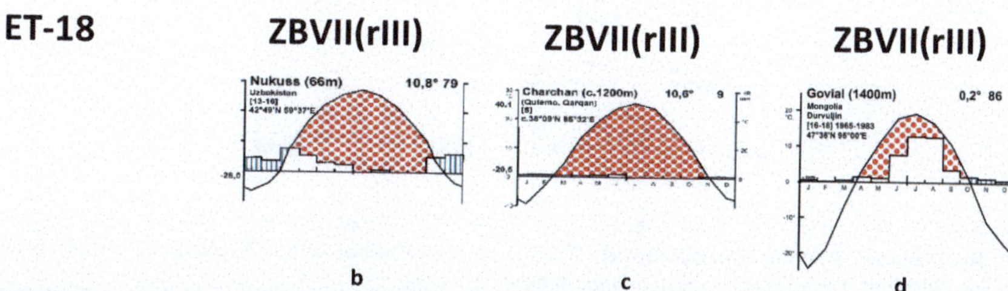

ET-18 (b-d): Temperate: Mesotherm/Hyperarid -- Temperate deserts: Nukuss/Uzbek {ZBVII(rIII), Fig. b}; Charchan/China {ZBVII(rIII), Fig. c}; Govial/Mongolia {ZBVII(rIII), Fig. d}

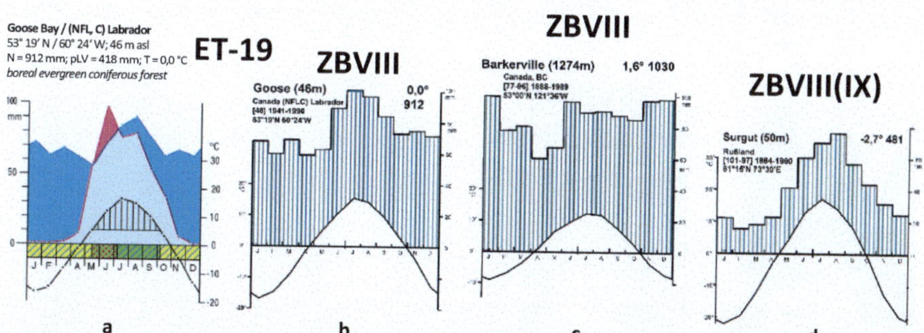

ET-19 (a-d): Boreal: Microtherm/Hyperhumid – Boreal conifer forests (Taiga): Goose Bay/Canada {ZBVIII(IX), Fig. a, b}; Barkerville/Canada {ZBVIII, Fig. c}; Surgut/Russia {ZBVIII(IX), Fig. d}

Fig. 4.10 (continued)

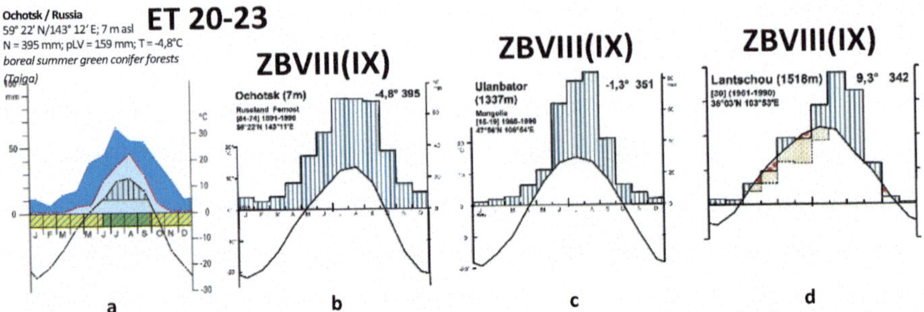

ET-20-23 (a-d): Boreal: Microtherm/Euhumid-Euarid -- Transition types of conifer forests and grass steppes (N-China, Mongolia, Tibet): Ochotsk/Russia, Far East {ZBVIII(IX), Fig. a, b}; Ulan-Batoor/Mongolia {ZBVIII(IX), Fig. c}; Lantschou/China {ZBVIII(IX), Fig. d}

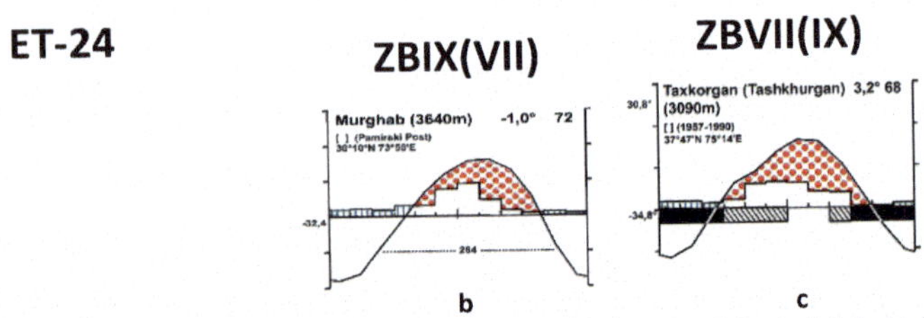

ET-24 (b-c): Boreal: Microtherm/Hyperarid – Cold deserts of Central Asia high plateaus: Murghab/Pamir/Tajik {ZBIX(VII), Fig. b}; Taxkorgan/Sinkiang {ZBVII(IX), Fig. c}

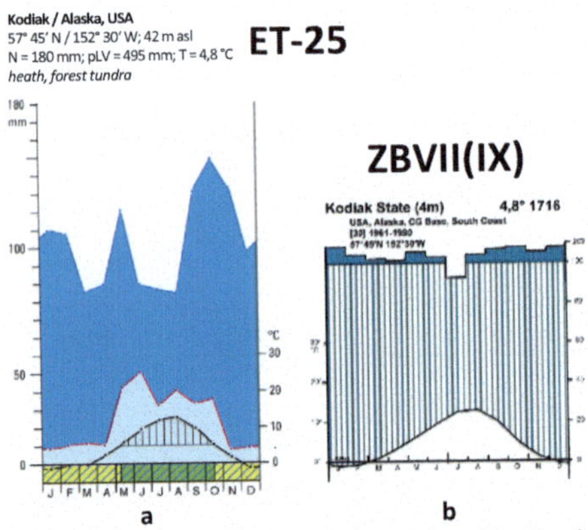

ET-25 (a-b): Polar: Oligotherm/Hyperhumid – Tundra: Kodiak/Alaska {ZBV(tIX), Fig. a, b}

Fig. 4.10 (continued)

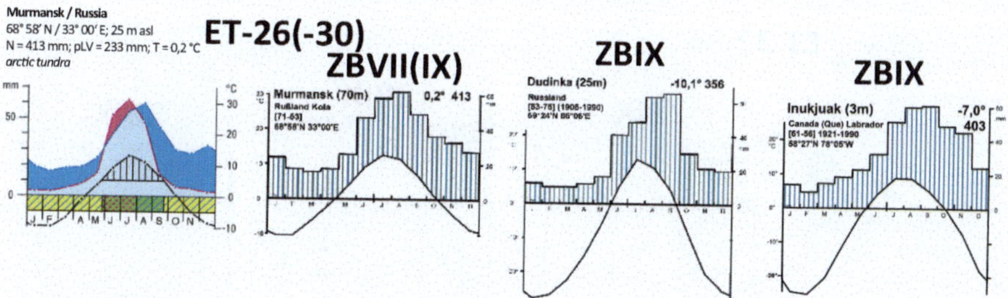

a b c d

ET-26(-30) (a-d): Polar: Oligotherm/Euhumid-Hyperarid – Different transition types of vegetation patches, forest tundra swamps, and frost debris in the subpolar regions and high altitudes all over the world: Murmansk/Russia {ZBVII(IX), Fig. a, b}; Dudinka/Russia {ZBIX, Fig. c}; Inukjuak/Canada {ZBIX, Fig. d}

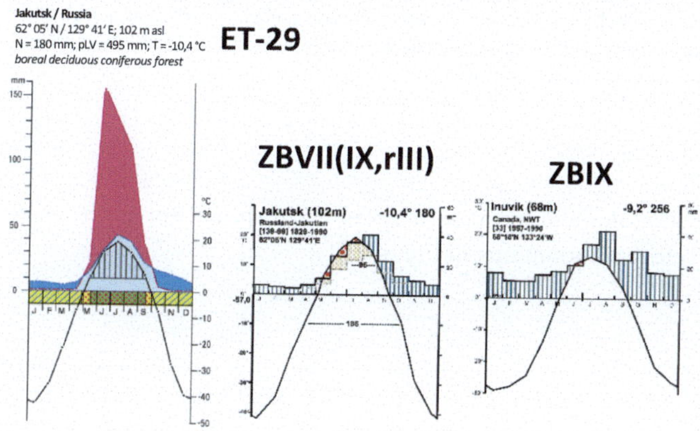

a b c

ET-29 (a-c): Polar: Oligotherm/Euarid – extrem continental, very dry steppes (Siberia): Jakutsk/Russia {ZBVII(IX,rIII), Fig. a, b}; Inuvik/Canada {ZBIX), Fig. c}

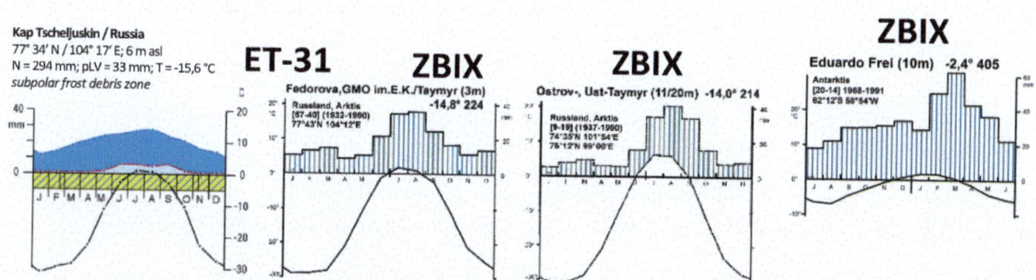

a b c d

ET-31 (a-c): Polar: Hekistotherm/Hyperhumid – Frost debris belt: Kap Tscheljuskin/Russia {ZBIX, Fig. a}, Fedorova/Russia {ZBIX, Fig. b}, Ostrov Ust Taymyr/Russia {ZBIX, Fig. c}, Eduardo Frei/Antarctica {ZBIX, Fig. d}

Fig. 4.10 (continued)

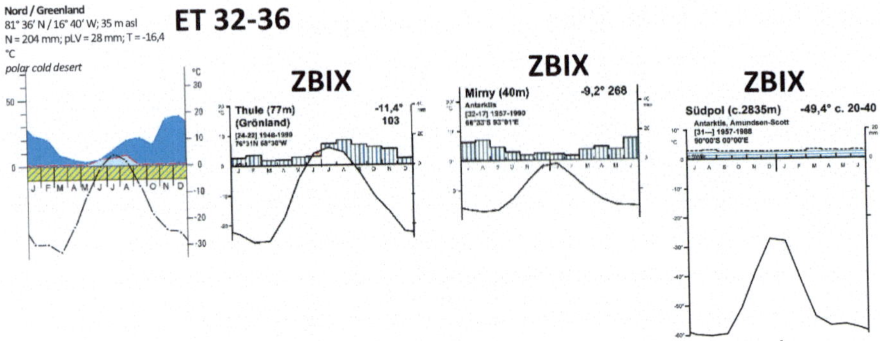

a b c d

ET-32-36 (a-c): Polar: Hekistotherm/Euhumid-Hyperarid – Polar ice caps and glaciated areas in high altitudes all over the world: North/Greenland {ZBIX, Fig. a}; Thule (North Greenland) {ZBIX, Fig. b}; Mirny/Antarctica {ZBIX, Fig. c}; South pole/Antarctica {ZBIX, Fig. d}

Fig. 4.10 (continued)

Corresponding typical ecological climatic diagrams are shown for each zonobiome in Fig. 4.11 and relevant mountains in Fig. 4.13.

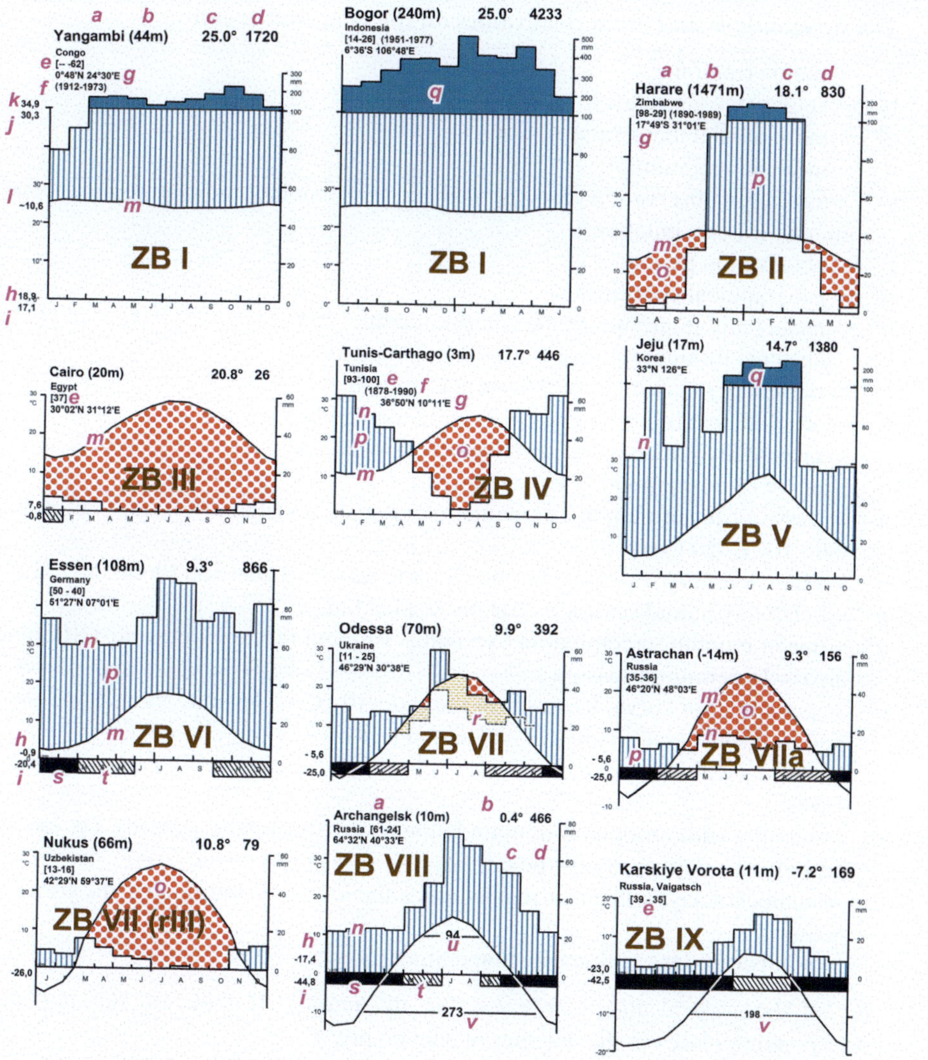

Fig. 4.11 Ecological climate diagrams by Walter and Lieth (1960–67), designed by Breckle (2002)

The designations and numerical values on the diagrams mean:

a = station (region);
b = height above sea level;
c = mean annual temperature;
d = mean annual rainfall;
e = number of observation years (possibly first number for temperature and second number for precipitation);
f = observation period;
g = geographical coordinates;
h = mean daily minimum of the coldest month;
i = absolute minimum (lowest measured temperature);
j = mean daily maximum of the warmest month;
k = absolute maximum (highest measured temperature);
l = mean daily temperature variation (for tropical stations).
m = curve of mean monthly temperatures;
n = curve of mean monthly precipitation (1 scale division = 20 mm, i.e. in the ratio 10 °C–20 mm);
o = relative aridity (drought) for the climatic region in question (dotted in red);
p = relatively humid season (vertically shaded blue);
q = mean monthly precipitation exceeding 100 mm (scale reduced to 1/10, blue area = hyperhumid season;
r = precipitation curve, lowered, in the ratio 10 °C = 30 mm, above horizontal yellow dashed area = relative dry season (only for steppe stations);
s = months with mean daily minimum below 0 °C (black) = cold season; periodic frosts
t = months with absolute minimum below 0 °C (diagonally shaded), i.e. late or early frosts may occur; episodic frosts
u = number of days with mean temperatures above +10 °C (duration of the growing season);
v = number of days with mean temperatures above −10 °C.

Not all data are available or given for all stations. If they are missing, the corresponding places in the diagram remain empty.

The climate diagrams see Fig. 4.10 and see Fig. 4.12 are particularly suitable for finding out **"homoclimates"**, which is extremely tedious when using extensive climate tables. One only needs to compare the given climate diagram with those in the Climate Diagram World Atlas from areas where homoclimates are suspected. Figure 4.12 shows, for example, the homoclimates of Karachi (Pakistan) from zonobiome III (dry slight transition to ZB II) and of Bombay (wet savannahs and forests, very typical of ZB II) in another part of the world. Knowledge of homoclimates is e.g. very important for new crop introductions in areas where they are not yet known. Another impressive example is the mediterranean climate with a dry summer and winter rains (cold season rains, ZB IV). This is shown (row bottom, Fig. 4.12), where the Mediterranean (the largest region), the Californian, the Middle Chilean, the South-African and the South resp. South-East Australian climate is characterized by the relevant climatic diagram. It is the opposite of the top and middle rows of Fig. 4.12, which show summer rains (warm-season rains, ZB II). The typical mediterranean climate is responsible for an amazing convergent evolution of sclerophyllic species from rather different taxonomic groups, thus forming a rather similar looking canopy (macchis, chaparral, matorral, fynbos, mallee; Walter and Breckle 1991).

If one looks at climate diagrams of mountain regions, then these are very similar to the diagrams of the surrounding lowlands over the course of the year; however, the temperature curve is more or less lowered depending on the elevation, the precipitation can often be increased. As a result, the aridity usually decreases significantly with increasing altitude. Corresponding mountain stations of the respective zonobiomes are shown as examples in Fig. 4.13.

To get a quick overview of the climate structure of larger areas, one can use climate diagram maps. These are obtained by inserting the ecological climate diagrams in the geographically correct position e.g. on large wall maps of individual continents or countries. The clarity is increased if the drought period in the diagram is marked in red and that of the humid periods in blue. Then the division can be overlooked at a glance. Such climate diagram maps of all continents in large format (black and white) have been published elsewhere (Walter et al. 1975). Clusters of similar climates and distinct gradients can be easily detected. In Fig. 4.14, as an example, the climate diagram map of Africa in small format with a few climate diagrams is shown.

The Climate Diagram World Atlas of Walter and Lieth (1960–67) contains over 8,000 diagrams. Climatic diagram maps of many selected regions with new designed climatic diagrams are also available at www.s.breckle.name/KlimaDiagrKarten/. The corresponding map showing the arrangement of zonobiomes and zonoecotones in Africa is given in Fig. 4.15.

The sequence of zonobiomes is arranged on both sides of the equator, almost symmetrically in Africa, but less so on other continents, modified by mountains and because the land masses in the southern hemisphere are smaller and the climate is more oceanic and cooler. The transition areas between the zonobiomes are the zonoecotones (see Fig. 4.15). Note that zonobiomes VI through IX are small in the southern hemisphere. Zonobiomes VI and VII are weakly developed in the southern hemisphere, ZB VIII is completely absent, and ZB IX is represented only by the subantarctic islands and the southern tip of South America, if one ignores the icy and nearly vegetationless Antarctic. The shift of boundaries in coming decades is shown in part II with the ecotypic climate classification.

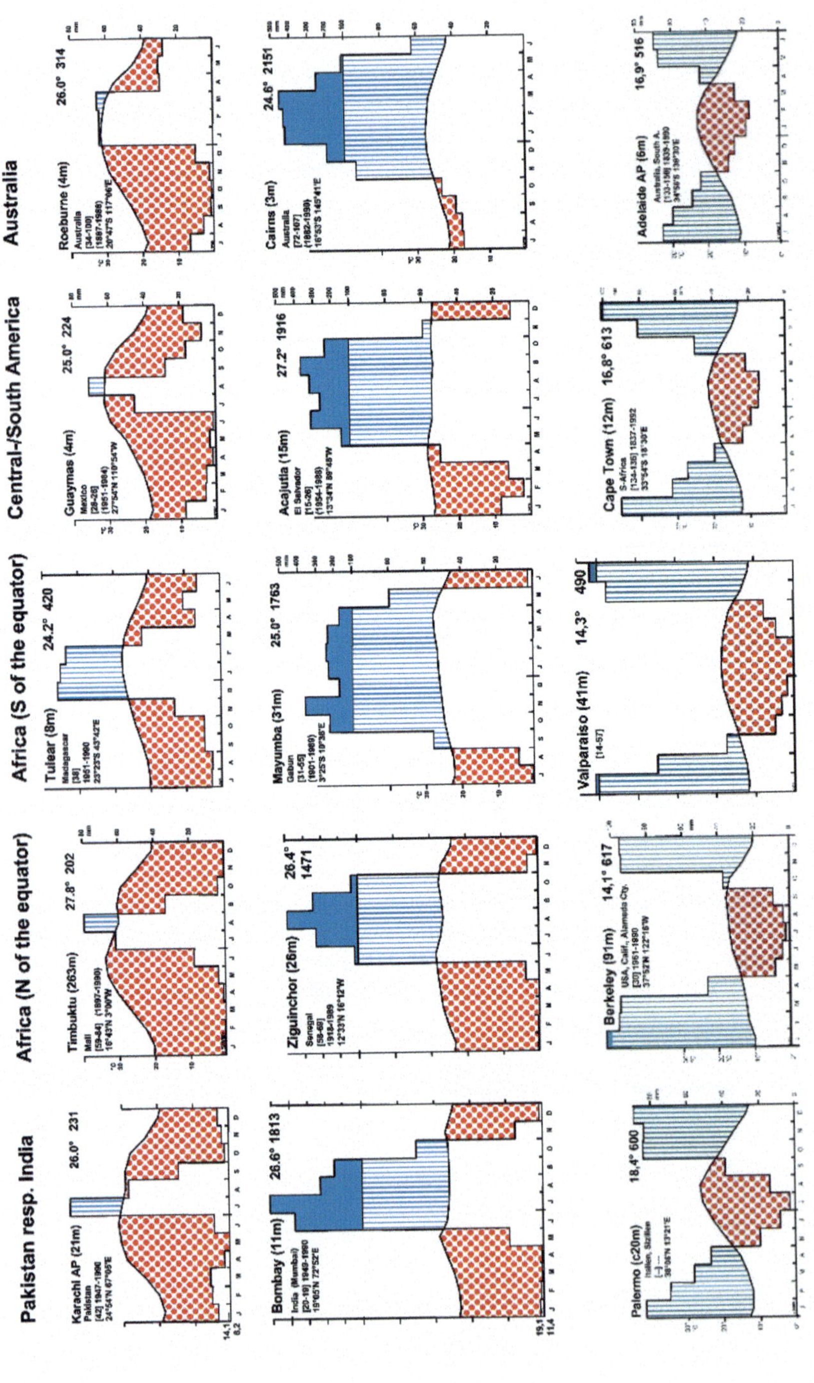

Fig. 4.12 The climate diagrams according to Walter and Lieth (1960–67) are good aids for identifying the *homoclimates*

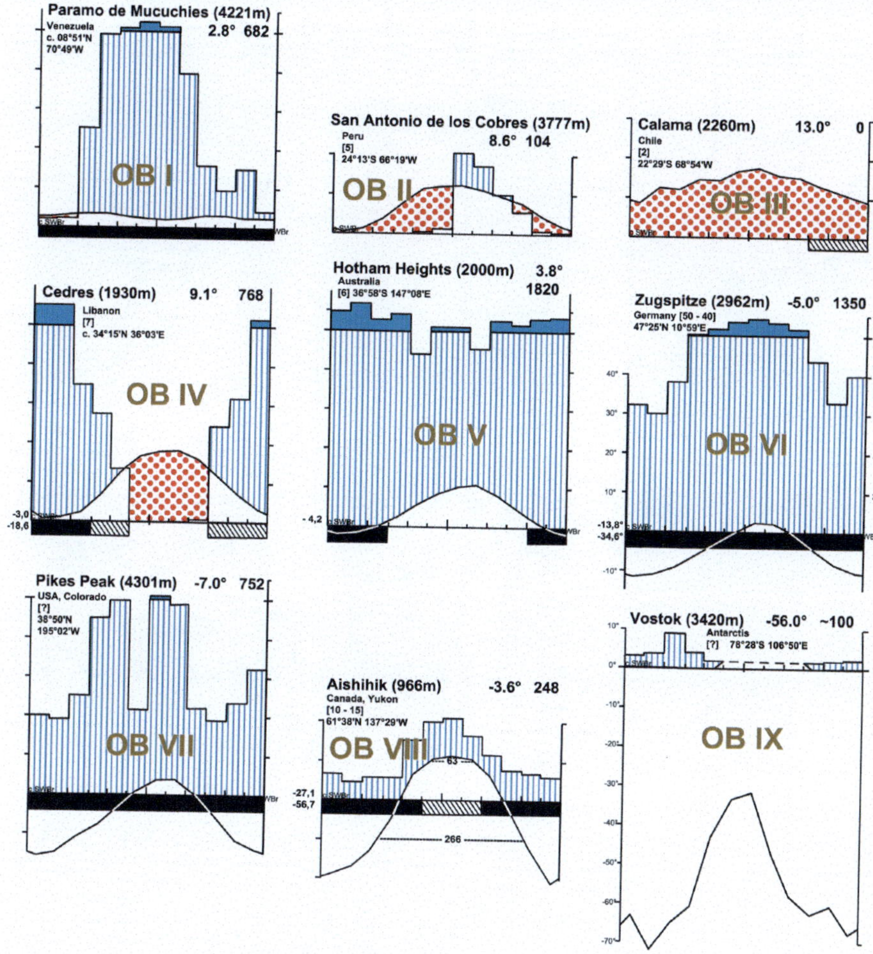

Fig. 4.13 Climate diagrams of the high mountains

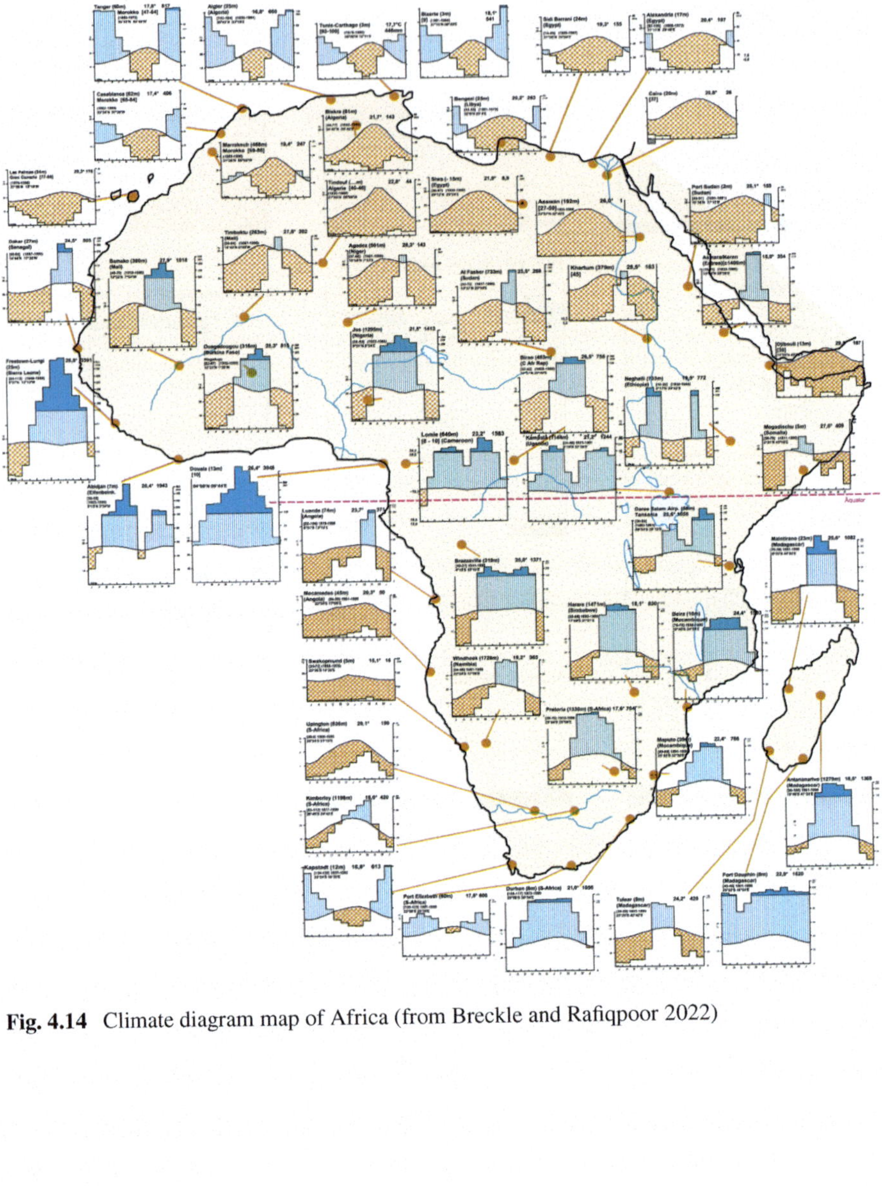

Fig. 4.14 Climate diagram map of Africa (from Breckle and Rafiqpoor 2022)

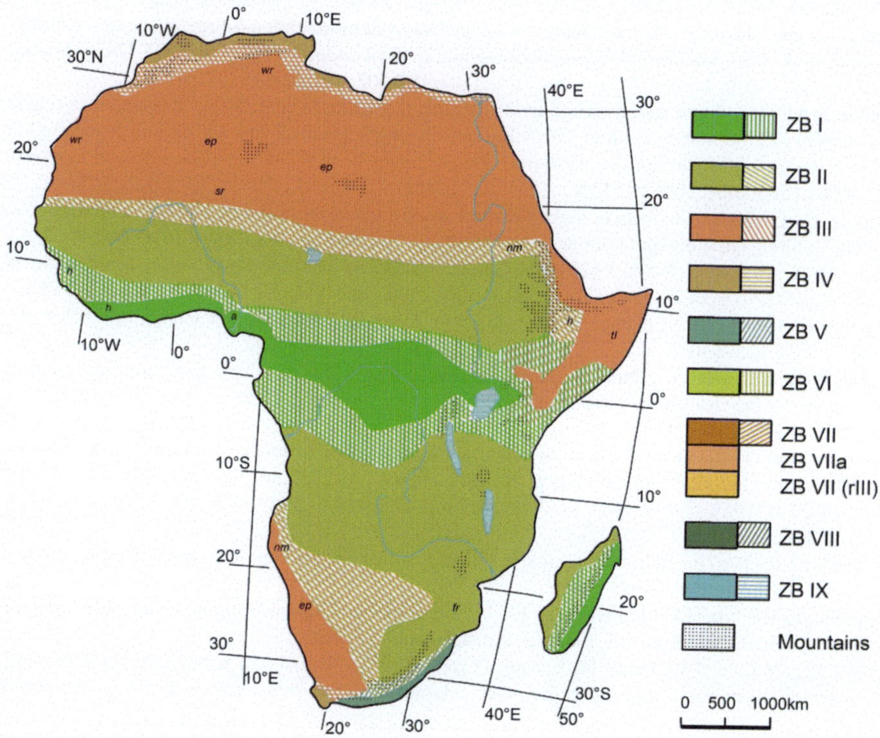

Fig. 4.15 The arrangement of zonobiomes and zonoecotones in Africa (from Breckle and Rafiqpoor 2022)

References

Allen RG, Pereira LS, Raes D, Smith M (1998) Crop evapotranspiration—guidelines for computing crop water requirements—FAO Irrigation and drainage paper 56. FAO—Food and Agriculture Organization of the United Nations Rome. ISBN: 92-5-104219-5. https://www.fao.org/4/X0490E/x0490e00.htm#Contents

Barthlott W, Mutke J, Rafiqpoor MD, Kier G, Kreft H (2005) Global centres of vascular plant diversity. Nova Acta Leopoldina, vol 92: 61–83. https://www.researchgate.net/publication/215672852_Global_centers_of_vascular_plant_diversity

Bendix J, Rafiqpoor MD (2001) Studies on the thermal conditions of soils at the upper tree line in the Páramo of Papallacta (Eastern Cordillera of Ecuador). Erdkunde 55:257–276

Breckle S-W (2002) Walter's Vegetation of the earth. The ecological systems of the Geo–Biosphere. Springer, Berlin

Breckle S-W, Rafiqpoor MD (2022) Vegetation and climate. Springer, Heidelberg, New York. ISBN: 789-3-662-64035-7

Brinkman WAR (1979) Growing season length as an indicator of climatic variations. Clim Change 2:127–138

Brown JA (1976) Shortening of the growing season length in the U.S. corn belt. Nature 260:421–421

Ellenberg, H, Mayer R, Schauermann J (1986) Ökosystemforschung—Ergebnisse des Solling-Projektes. Zeitschrift für Pflanzenernährung und Bodenkunde, vol 151(4): pp 507. Ulmer Verlag, Stuttgart. ISBN: 3-8001-3431-4. https://doi.org/10.1002/jpln.19881510415
Euskirchen ES, McGuire AD, Kicklighter DW, Zhuang Q, Clein JS et al (2006) Importance of recent shifts in soil thermal dynamics on growing season length, productivity, and carbon sequestration in terrestrial high–latitude ecosystems. Glob Change Biol 12(4):731–750. https://doi.org/10.1111/j.1365-2486.2006.01113.x
Feng S, Hu Q (2004) Changes in agro-meteorological indicators in the contiguous United States: 1951–2000. Theoret Appl Climatol 78:247–264. https://doi.org/10.1007/s00704-004-0061-8
Fisher JB, Whittaker RJ, Malhi Y (2010) Et come home: potential evapotranspiration in geographical ecology. Glob Ecol Biogeogr 20(1):1–18. https://doi.org/10.1111/j.1466-8238.2010.00578.x
Gaussen H (1954) Théorie et classification des climats et microclimates, 8me Congr. Internat. Paris, sect. 7+3, 125–130
Hijmans RJ, Cameron SE, Parra JL, Jones PG, Jarvis A (2005) Very high-resolution interpolated climate surface for global land areas. Int J Climatol 25:1965–1978. https://doi.org/10.1002/joc.1276
Holdridge LR (1947) Determination of world plant formations from simple climatic data. Science 105(2727): 367–368. https://doi.org/10.1126/science.105.2727.367
Holdridge LR (1959) Simple method for determining potential evapotranspiration from temperature data. Science 130(3375):572. https://doi.org/10.1126/science.130.3375.572.a
Holdridge LR (1967): Life zone ecology. Tropical Science Centre, San Jose. https://app.ingemmet.gob.pe/biblioteca/pdf/Amb-56.pdf
Hutchinson MF, Nix HA, Mcmahon JP (1992) Climate constraints on cropping systems. In: Pearson CJ (ed) Field Crop Systems. Elsevier, Amsterdam, pp 37–58
Lozan JL, Breckle S-W, Graßl H, Kasang D (eds) (2021) Warnsignal Klima: Boden und Landnutzung. Wissensch. Auswertungen, Geo, Hamburg. https://www.fdr.uni-hamburg.de/record/9905
Johnson AD (2006) Resource allocation in plants. Association for Biology Laboratory Education (ABLE) 2005 Proceedings, vol 27: 15–30. https://www.ableweb.org/biologylabs/wp-content/uploads/volumes/vol-27/02_Johnson.pdf
Junghans H (1969) Sonnenscheindauer und Strahlungsempfang geneigter Ebenen. Abhandlungen des Meteorologischen Dienstes der DDR, Nr. 58(11), Berlin
Kessler A (1985) Anthropogene Änderungen des Strahlungshaushaltes der Erdoberfläche. Erdkunde 39:175–179
Köppen W (1900) Versuch einer Klassifikation der Klimate, vorzugsweise nach ihren Beziehungen zur Pflanzenwelt. Geographische Zeitschrift, vol 6: 593–611 und 657–679
Körner Ch (2007) Climatic treelines: conventions, global patterns, causes. Erdkunde 61:316–324
Kottek M, Grieser M, Beck Ch, Rudolf B, Rubel F (2006) World map of the Köppen-Geiger climate classification updated. Meteorol Z 15:259–263
Kunkel KE, Easterling DR, Refvdmond KT, Hubbard KG (2004) Temporal variations in frost-free season in the United States: 1895–2000. Geophys Res Lett 31(3):1–4
Larcher W (1980) Ökologie der Pflanzen. UTB, Stuttgart, p 232
Latrubesse EM, Baker PA, Argollo J (2009) Geomorphology of natural hazards and human-induced disasters in Bolivia. Deve Earth Surface Process 13:181–194. https://doi.org/10.1016/S0928-2025(08)10010-4
Lauer W (1952) Humide und aride Jahreszeiten in Afrika und Südamerika und ihre Beziehungen zu den Vegetationsgürteln. Bonner Geographische Abhandlungen 9:15–98
Lauer W (1991) Climatology. In: Pancel L (ed) Tropical forestry handbook. Springer Verlag, Berlin, Heidelberg, New York, pp 95–164
Lauer W, Frankenberg P (1981a) Untersuchungen zur Humidität und Aridität von Afrika. Das Konzept einer potentiellen Landschaftsverdunstung. Bonner Geographische Abhandlungen Bd. 66

Lauer W, Frankenberg P (1981b) Eine Karte der hygrothermischen Klimatypen Afrikas. Erdkunde 35:245–248

Lauer W, Frankenberg P (1982) Weltkarten der Relation von fühlbarer und latenter Wärme. Erdkunde 36:137–142

Lauer W, Klaus D (1975) The thermal circulation of the central Mexican Meseta region within influence of the trade winds. Arch. für Meteorologie Geophysik und Bioklimatologie. Serie B 23:342–366

Lauer W, Rafiqpoor MD (2002) Die Klimate der Erde—eine Klassifikation auf der Grundlage der ökophysiologischen Merkmale der realen Vegetation. Franz Steiner Verlag, Stuttgart, p 271

Lauer W, Rafiqpoor MD, Frankenberg P (1996) Die Klimate der Erde—eine Klassifikation auf ökophysiologischer Grundlage der realen Vegetation. Erdkunde 50:257–300

Leemans R, Crame W, Minnen JGV (1996) Prediction of global biome distribution using bioclimatic equilibrium models. In: Breymeyer, Hall DO, Melillo JM, Ågren GI (eds) Global change: Effects on coniferous forests and grasslands. John Wiley & Sons Ltd, New York

Louis H (1958) Der Bestrahlungsgang als Fundamentalerscheinung der geographischen Klimaunterscheidung. Geographische Forschung. In: Paschiner H. (ed) Festschrift für Hans Kinzl. Schlern-Schriften, pp 155–164

Miller P, Mitchel M (2005) Climate Change: Length of growing season in the U.S. corn belt. Phys Geogr 26:85–98

Mitchell TD, Hulme M (2002) Length of the growing season. Weather 57:196–198

Moran J, Morgan MD (1977) Recent trends in hemispheric temperature and growing season indices in Wisconsin. Agric Meteorol 18:1–8

Paulsen J, Körner Ch (2014) A climate-based model to predict potential treeline position around the globe. Alp Bot 124:1–12. https://doi.org/10.1007/s00035-014-0124-0-

Penck A (1910) Versuch einer Klimaklassifikation auf physiographischer Grundlage. Sitzungsberichte der Preußischen Akademie der Wissenschaften, Physisch Mathematische Klasse 12:236–246

Penman HL (1948) Natural evaporation from open water, bare soil and grass. Proceedings of the royal society series a—mathematical, physical and engineering sciences, vol 193(1032): 120–145. https://doi.org/10.1098/rspa.1948.0037

Pezza AB, Ambrizzi T (2005) Dynamical conditions and synoptic tracks associated with different types of cold surges over Tropical South America. Int J Climatol 25(2):215–241. https://doi.org/10.1002/joc.1080

Rafiqpoor MD (2024): Climatology. اقلیم شناسی—DCM-Druck Center Meckenheim. Germany, pp 298. https://www.researchgate.net/publication/377066161_Climatology_-_aqlym_shnasy

Sharratt BS (1992) Growing season trends in the Alaskan climate record. Arctic 45:124–127

Stern H, De Hoedt G, Ernst J (2000) Objective classification of Australian climate. Aust. Meteor. Mag., vol 49(2): 87–96. https://www.researchgate.net/publication/279766565_Objective_classification_of_Australian_climates

Thornthwaite CW (1943) Problems in the classification of climates. Geogr Rev 33(2):233–255

Thornthwaite CW (1948) An approach toward a rational classification of climate. Geogr Rev 38:55–94

Treter U (1993) Die borealen Waldländer. Westermann Verlag, Braunschweig, p 210

Tuhkanen S (1984) A circumboreal system of climatic–phytogeographical regions. Acta Botanica Finica, vol 127. ISBN: 951-9469-21-4

Walter H (1954) Einführung in die Phytologie. Band III: Grundlage der Pflanzenverbreitung, Teil I: Standortlehre, Stuttgart

Walter H (1979) Die ökologischen Systeme der Kontinente (Biosphäre). Stuttgart pp 131

Walter H, Breckle S-W (1991) Ökologie der Erde Band 1—Grundlagen. 2. Auflage. Fischer Verlag, Stuttgart

Walter H, Lieth H (1960–67): Klimadiagramm-Weltatlas. VEB Gustav Fischer Verlag Jena

Walter H, Harnickel, E, Mueller-Dombois D (1975) Klimadiagrammkarten der einzelnen Kontinente und ökologische Klimagliederung der Kontinente. Fischer/Stuttgart, pp 36

Whittaker RH (1975) Communities and ecosystems. Macmillan, New York

Woodward FI, Lomas MR (2004) Vegetation dynamics: Simulating responses to climatic change. Biol Rev Camb Phil Soc 79:643–670. https://doi.org/10.1017/s1464793103006419

Woodward FI, Lomas MR, Kelly CK (2004) Global climate and the distribution of plant biomes. Proc Royal Soc London Series B Biol Sci 359:1465–1476. https://doi.org/10.1098/rstb.2004.1525

Woodward FI, Smith TM, Emanuel WR (1995) A global land primary productivity and phytogeography model. Global Biogeoch Cycles 9:471–490.https://doi.org/10.1029/95GB02432

Zepner L, Karrassch P, Wiemann F, Bernard L (2020) ClimateCharts.net—an interactive climate analysis web platform. Int J Digital Earth 14(3). https://doi.org/10.1080/17538947.2020.1829112

Zomer R, Trabucco A (2022) global aridity index and potential evapotranspiration climate database v3—algorithm code (Python). https://doi.org/10.6084/m9.figshare.20005589

Chapter 5
Summary—Climate and Biodiversity

The ecoclimate classification presented here differs from a previous study by Lauer and Rafiqpoor (2002) in that it provides a quantitative fully reproducible delineation of ecoclimatic boundaries. The variables used to classify the world's climates are ecophysiologically meaningful parameters (Box 2004) and provide a more accurate picture than annual averages (Holdridge 1947, 1967; Whittaker 1975; Leemans et al. 1996; Leser 2005). Our ecoclimate classification is based on the most important climatic factors for plant life, namely the availability of energy (isothermomenes) and water (isohygromenes), which limit or promote the ecophysiological requirements of plants and the duration of the photosynthetically active period of the year (O'brien 2006). Despite using few and simple climatic parameters and a small number of assumptions, the ecoclimate classification is able to capture the distribution of the world's major biomes (Walter and Breckle 1985; Loidi et al. 2022).

The ecoclimate classification can provide a basis for the analysis of species richness gradients (see Fig. 4.3D). Most of the centres of plant richness fall into the megatherme-hyperhumid climate type (cf. Barthlott et al. 2005; Fig. 5.1). All of the top five maxima of species richness (>5,000 spp. per 10,000 km^2: Barthlott et al. 2005, 2014) are located in mountainous regions within the megatherm-hyperhumid tropics (Fig. 5.2). Here, the hygrothermal conditions in combination with other factors of geodiversity (e.g., relief, geology, soil) (Lauer and Rafiqpoor 2001) are favourable (see Fig. 4.8) for maintaining the highest number of plant species. In addition, the ecoclimatic diversity usually associated with mountainous regions increases the potential for the coexistence of many plant species at the spatial scale studied (Barthlott et al. 2005, 2007, Barthlott et al., 2014; Mutke and Barthlott 2005). Areas with minimal species richness (<200 species per 10,000 km^2) fall into either hyperarid subtropical and tropical deserts or hekistothermal Arctic or high mountain deserts (see Figs. 5.1 and 5.2). The increase in the duration of the hygrothermal growing season (see Fig. 4.3C) is paralleled by the latitudinal gradient of species richness per latitudinal band (see Fig. 4.3E).

M. D. Rafiqpoor and S.-W. Breckle, *The World's Ecoclimates*,
SpringerBriefs in Earth System Sciences,
https://doi.org/10.1007/978-3-032-06456-1_5

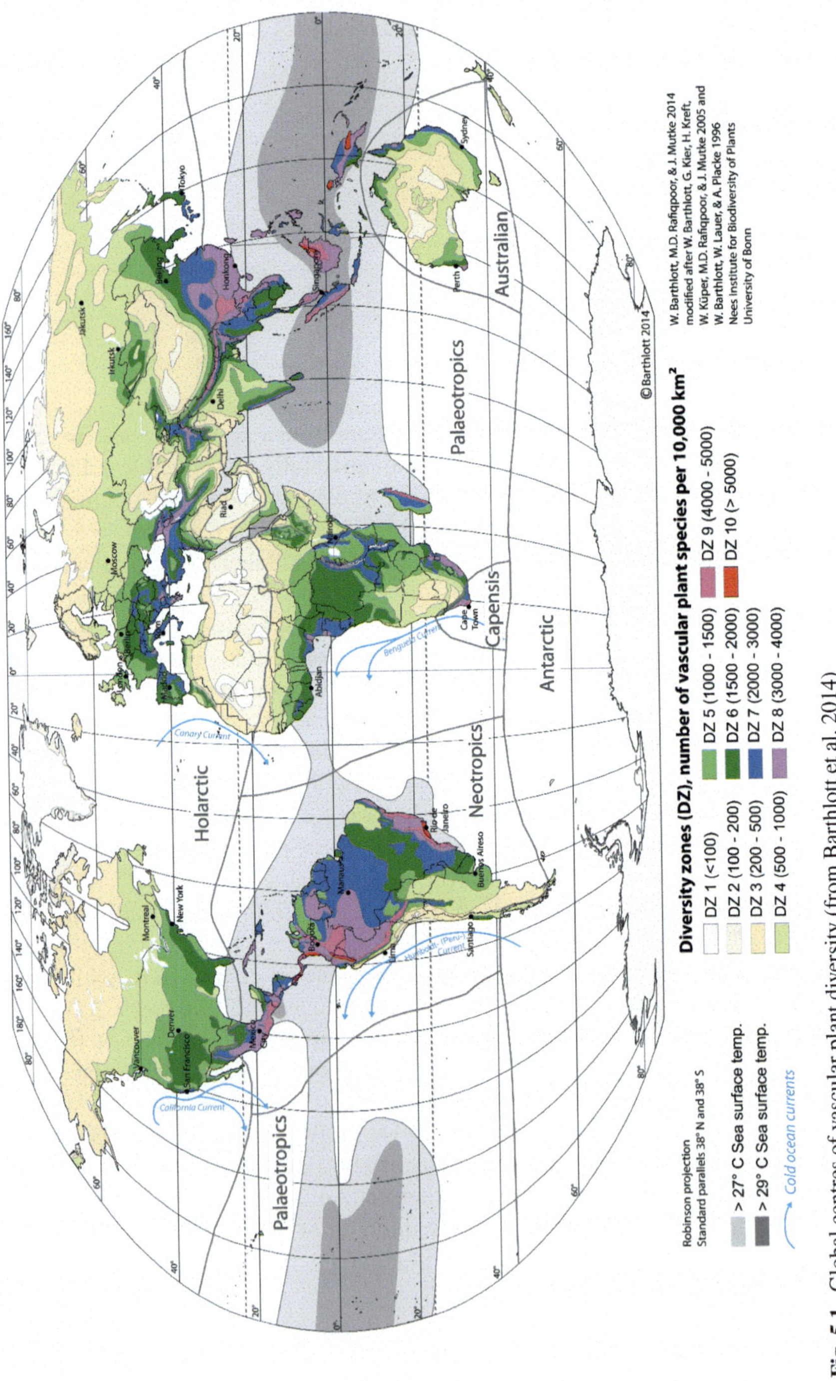

Fig. 5.1 Global centres of vascular plant diversity (from Barthlott et al. 2014)

CLIMATIC ZONES	Heat budget: number of months allowing thermal plant growth		Water budget: number of humid months allowing hygric plant growth: hyperhumid (12, 11, 10)	euhumid (9, 8, 7)	subhumid (6, 5)	semiarid (4, 3)	euarid (2, 1)	hyperarid (0)
A TROPIC (23,5° N and S)	Megatherm (in all climatic belts)	12	1154	915	542	267	101	22
B SUBTROPIC (45° N and S)	Macrotherm	11, 10, 9, 8, 7	715	342	217	137	108	31
C TEMPERATE (57° N and S)	Mesotherm	6, 5	449	280	140	225	105	14
D BOREAL (66,6° N and S)	Microtherm	4, 3	291	225	153	139	69	23
E POLAR	Oligotherm	2, 1	184	162	132	99	98	44
glaciated areas in A, B, C, D, E	Hekistotherm (in all climate zones)	0	76	57	83	66	50	14

Fig. 5.2 Species richness of plants per 10,000 km^2 (median of diversity zones calculated after data from Barthlott et al. 2005) in the corresponding ecoclimatic regions

References

Barthlott W, Mutke J, Rafiqpoor MD, Kier G, Kreft H (2005) Global centres of vascular plant diversity. Nova Acta Leopoldina 92:61–83. https://www.researchgate.net/publication/215672852_Global_centers_of_vascular_plant_diversity

Barthlott W, Hostert A, Kier G, Küper W, Kreft H, Mutke J, Rafiqpoor MD, Sommer JH (2007) Geographic patterns of plant diversity at continental to global scales. Erdkunde 61:305–316. https://www.researchgate.net/publication/215672847_Geographic_patterns_of_vascular_plant_diversity_at_continental_to_global_scale

Barthlott W, Erdelen RW & Rafiqpoor MD (2014) Biodiversity and technical innovations: bionics. In: Lanzerath D, Friele M (eds) Concept and value in biodiversity. routledge studies in biodiversity politics and management. pp 300–315. ISBN: 978-1-415-66057-0. https://www.routledge.com/Concepts-and-Values-in-Biodiversity/Lanzerath-Friele/p/book/9780415660570

Box EO (2004) Review to the book of Lauer W & Rafiqpoor MD (2002) Die Klimate der Erde: Eine Klassifikation auf der Grundlage der ökophysiologischen Merkmale der realen Vegetation. Phytocoenologia 34(4):673–675. https://www.schweizerbart.de/papers/phyto/list/34#issue1

Holdridge LR (1947) Determination of World Plant formations from simple climatic data. Science 105(2727):367–368. https://doi.org/10.1126/science.105.2727.367

Holdridge LR (1967) Life zone ecology. Tropical Science Centre, San Jose. https://app.ingemmet.gob.pe/biblioteca/pdf/Amb-56.pdf

Lauer W, Rafiqpoor MD (2001) Biodiversität - Geodiversität - Ökodiversität. Die Artenvielfalt – der lebendige Reichtum der Erde. In: Meyers Großer Weltatlas. Atlas zum 21. Jahrhundert. Meyers Lexikonverlag, Mannheim, pp 28–29

Lauer W, Rafiqpoor MD (2002) Die Klimate der Erde – eine Klassifikation auf der Grundlage der ökophysiologischen Merkmale der realen Vegetation. Franz Steiner Verlag, Stuttgart, p 271

Leemans R, Crame W, Minnen JGV (1996) Prediction of global biome distribution using bioclimatic equilibrium models. In: Breymeyer, Hall DO, Melillo JM, Ågren GI (eds) Global change: effects on coniferous forests and grasslands. John Wiley & Sons Ltd, New York

Leser H (2005) Review to the book of Lauer W. & Rafiqpoor MD "Die Klimate der Erde - Eine Klassifikation auf der Grundlage der ökophysiologischen Merkmale der realen Vegetation". Geographica Helvetica 60:68–69

Loidi J, Navarro-Sánchez G, Vynokurov D (2022) Climatic definitions of the world's terrestrial biomes. Vegetation Classification and Survey 3:231–271. https://www.researchgate.net/publication/366408169_Climatic_definitions_of_the_world's_terrestrial_biomes

Mutke J, Barthlott W (2005) Patterns of vascular plant diversity at continental to global scales. Biol Skr 55:521–531

O'brien EM (2006) Biological relativity to water–energy dynamics. J Biogeogr 33:1868–1888

Walter H, Breckle S-W (1985) Vegetation of the earth and ecological systems of the Geobiosphere. 3rd ed. Springer, Berlin, pp 319. https://doi.org/10.1007/978-3-662-02437-9_3

Whittaker RH (1975) Communities and ecosystems. Macmillan, New York

Part II
Ecoclimate Classification—A Powerful Tool for Modelling Future Climate Change Scenarios

With the help of high-resolution data on temperature, precipitation and potential evapotranspiration, an ecoclimate classification for the middle (2050) and the end of the 21st century (2090) is created for the IPCC **B1** and IPCC SRES **A2** scenarios on the basis of coupled atmosphere-ocean **G**eneral **C**irculation **M**odels (**GCMs**). By the end of the 21st century, there will be major shifts in climate zones, especially in the high latitudes of the northern hemisphere, by almost 5–6 degrees of latitude polewards. Central Europe (favoured by hygrothermal conditions) and southern Europe (increase in aridity) experience strong spatial changes in climate types. In the subtropics, the spatial variation of **latitudinal** and **elevational** shifts of **climate zones** and **climate belts** is highest due to the dominance of young mountain ranges and thus high habitat heterogeneity ('Geodiversity': Lauer & Rafiqpoor 2001). In the marginal tropics, megatherme tropical climate types are shifting worldwide towards the subtropics of both hemispheres. Minor spatial shifts in climate zones occur in the tropical lowlands. The tropical mountains, however, are experiencing dramatic altitudinal shifts in elevational belts worldwide: The glaciated high mountain regions are almost disappearing. In Asia, the monsoon circulation is intensifying due to the intensification of the thermal land/water contrast combined with hygrothermal climate favouritism, especially in China, Japan and Korea. The results of our investigations are confirmed by findings from small-scale climate modelling and 'climate envelope' studies for individual parts of the world. The classification thus shows a global synthesis of climate zone shifts for the selected time slices of the 21st century on the basis of climatic household variables of ecological relevance.

Reference

Lauer W, Rafiqpoor MD (2001) Biodiversität—Geodiversität—Ökodiversität. Die Artenvielfalt—der lebendige Reichtum der Erde. In: Meyers Großer Weltatlas. Atlas zum 21. Jahrhundert. Meyers Lexikonverlag, Mannheim, pp 28–29

Chapter 6
Introduction

In February 2007, the "Contribution of Working Group I to the **F**ourth **A**ssessment **R**eport" (**FAR** or **AR4**) of the **I**ntergovernmental **P**anel on **C**limate **C**hange (**IPCC**) was published, providing detailed descriptions of the IPCC **S**pecial **R**eport on **E**missions **S**cenarios (**SRES**). In the **AR4**, the IPCC had presented six scenarios of future emissions based on very different assumptions about future population and economic growth, energy consumption, technological developments and land use. Since then, six further **A**ssessment **R**eport (**AR6** released on 20 March 2023) have been published up to date.

Since 2005, the **C**limate **C**hange **P**erformance **I**ndex (**CCPI**) has been released. The CCPI uses a standardized framework to compare the climate performance of 63 countries and the EU, which together account for over 90% of global greenhouse gas (GHG) emissions (EU). The climate mitigation performance is assessed in four categories: GHG Emissions, Renewable Energy, Energy Use and Climate Policy. The **CCPI 2025** has been published in November 20th 2024. The ranking results are depicted on a map (Fig. 6.1).

The CCPI-index improves transparency in international climate policy, enabling a comparison of the climate protection performance and progress of individual countries. In addition, COP conferences have been held regularly every year since 1995, at which the participating countries discuss ways of tackling climate change and sustainably preserving the world for future generations—with very varied and unsatisfactory results to date (https://de.wikipedia.org/wiki/UN-Klimakonferenz). The last COP29 took place in Baku in November 11th to 24th 2024. Like the former COP-Conferences, "*the current COP29 in Baku is not delivering what would actually have been necessary—but it is at the upper end of what is possible in the current political climate*", says Christoph Bals, Political Director of Germanwatch (https://www.germanwatch.org/de/91798). However, it now seems to be recognised that the rich industrialised countries, which emit the most greenhouse gases, are the main causers of global warming, but that the main consequences will have to be borne

M. D. Rafiqpoor and S.-W. Breckle, *The World's Ecoclimates*,
SpringerBriefs in Earth System Sciences,
https://doi.org/10.1007/978-3-032-06456-1_6

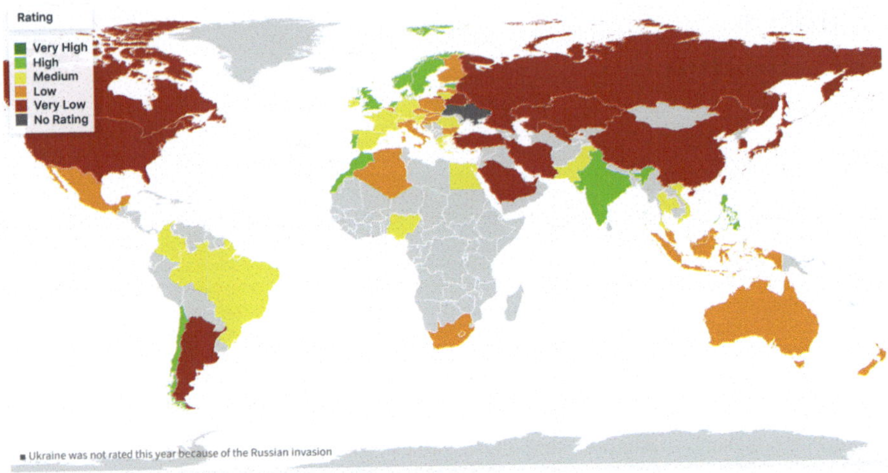

Fig. 6.1 Results of the CCPI 2025 Ranking. "*The climate crisis is an existential threat to life on Earth. To reduce the magnitude of the crisis' impacts, we must limit global warming to 1.5 K, as decided in the Paris Agreement. Only decisive action will reduce greenhouse gas emissions, which are responsible for climate change.* ***As an independent monitoring tool, the CCPI has a leading role in informing on the Paris Agreement's implementation phase. Since 2005, the CCPI has provided analysis of countries' climate protection performance***" (from https://ccpi.org/). The goal will not be achieved

by poorer countries. (https://education.cfr.org/learn/reading/who-releases-most-greenhouse-gases). Furthermore, these GHG emitters are still doing nothing or very little to minimize GHG emissions in their countries (see Fig. 6.1). Figure 6.2 shows the contribution of each country to the change in global mean surface temperature (GMST, °C) resulting from historical emissions of GHGs (CO_2, CH_4 and N_2O) over three time periods. At COP29 it was therefore decided to increase aid for developing countries to combat the consequences of climate change. (https://www.germanwatch.org/de/91798). It remains to be hoped that the donor countries (i.e., the main causers of the GHG emissions) will meet their commitments and that the funds provided will not be channelled into the wrong channels in the recipient countries.

To return to the core of the problem, the IPCC-Reports, however, all confirm the rising global temperatures and spatial changes in the amount and distribution of precipitation (IPCC 2023). The temperature of the Earth's atmosphere has risen significantly over the past 100 years as a result of human activities, which have emitted greenhouse gases such as carbon dioxide (CO_2), methane (CH_4), nitrous oxide (N_2O), chlorofluorocarbons (CFCs), etc. These emissions have increased in recent decades, causing significant damage to nature, environment and the economies worldwide. To some extent, climate change has led to an accelerated increase in the melting of polar ice caps and high-mountain glaciers and a rise in sea levels. It has changed the temporal and spatial distribution of precipitation around the world, causing heavy rainfalls and floods, landslides and other disasters. Prolonged droughts

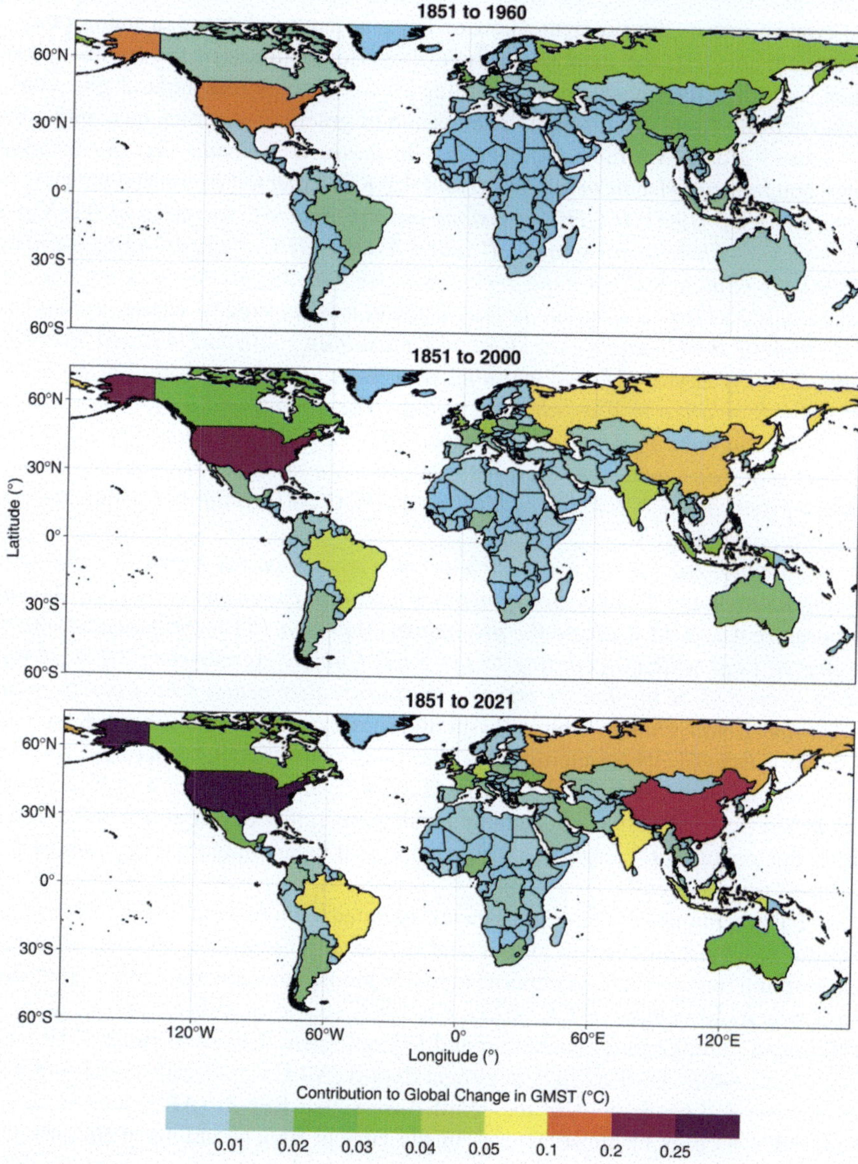

Fig. 6.2 National contributions to change **g**lobal **m**ean **s**urface **t**emperature (**GMST**, °C = K) resulting from historical emissions of CO_2, CH_4 and N_2O during three time periods. All data used in the map are provided in the **Data Records** at https://www.nature.com/articles/s41597-023-02041-1#ref-CR44 (map taken from: https://www.nature.com/articles/s41597-023-02041-1)

have contributed to the desertification of large agricultural areas in many places. Climate change, e.g. the increase in the average temperature of the Earth's atmosphere, is indirectly caused by the burning of fossil fuels (oil, natural gas, coal). Human activities, mainly through the emission of greenhouse gases, have undoubtedly caused global warming. "*Depending on future GHG emissions, an average global temperature rise of between 1.6 and 4.7 K is expected by 2100 compared to the period 1850–1900. If GHG emissions are not reduced, warming of 0.2 K per decade is very likely over the next 30 years. Even if GHG concentrations stabilize by 2100, the climate will continue to change beyond the twenty-first century and sea levels in particular will continue to rise. Current models suggest virtually complete elimination of the Greenland ice sheet and a resulting contribution to sea level rise of about 7 m if global average warming were sustained for millennia in excess of 1.9 to 4.6 K relative to pre-industrial values. The corresponding future temperatures in Greenland are comparable to those inferred for the last interglacial period 125,000 years ago, when paleoclimatic information suggests reductions of polar land ice extent and 4 to 6 m of sea level rise*. (IPCC 2007, *page 12)*" (https://www.ipcc.ch/report/ar4/syr/)

"*Meanwhile, 2023 is the warmest year since measurements began. According to the European Union's Copernicus Earth Observation Programme, the global mean temperature was 1.48 K above the pre-industrial period 1850–1900, just below the 1.5-degree limit which, according to the Paris Climate Agreement (2015), should not be exceeded in the longer term in the twenty-first century in order to avoid dangerous climate change*" (https://wiki.bildungsserver.de/klimawandel/index.php/Aktuelle_Klima%C3%A4nderungen). The EU climate service Copernicus sees the Paris climate targets already missed in 2024 with an average temperature rise of 1.55 K. Do we inevitably reach a global mean temperature of 3 K above present time within this century? (Wiegandt 2022), what would mean an increase by almost 6 K over land surface areas.

Global greenhouse gas emissions have continued to increase at historically uneven rates due to unsustainable energy and land use, as well as land use change, lifestyles, consumption and production patterns across regions, between and within countries, and between individuals (IPCC 2023). The IPCC scenarios thus cover a wide range of the main demographic, economic and technological drivers of future GHG and sulphur emissions. The Fourth Assessment Report (AR4) was already consistent with the analysis and projections of the Third Assessment Report (AR3), but the scope of the AR4 focuses on those aspects of the current understanding of the physical science of climate change that are judged to be most relevant to policy makers (IPCC AR4 2007; Crippa et al. 2022; IPCC AR6 2023: Summary for Policymakers: Doi: https://doi.org/10.59327/IPCC/AR6-9789291691647.001).

Sommer et al. (2010) analysed the relationships between contemporary water-energy dynamics and other non-climatic predictor variables to model the regional capacity for plant species richness (CSR) and its projected future changes. They found that under all analysed IPCC scenarios, relative changes in CSR increase with increasing projected temperature rise. By 2100, global average CSR is projected to remain similar to today (+0.3%) under the optimistic B1/ + 1.8 K scenario, but to decrease significantly (−9.4%) under the business-as-usual A1FI/ + 4.0 K scenario (Fig. 6.3). They summarized that: "*While in most temperate and arctic regions, a CSR increase is expected, the projections indicate a strong decline in most tropical and subtropical regions. Countries least responsible for past and present greenhouse gas emissions are likely to incur disproportionately large future losses in CSR, whereas industrialized countries have projected moderate increases.*

Independent of direction, we infer that all changes in regional CSR will probably induce on-site species turnover and thereby be a threat to native floras (Sommer et al. 2010). Figure 6.3 shows where biodiversity will decline globally by the end of the 21th century (especially in the tropics and their biodiversity centres), but that it will increase in temperate and subtropical areas such as Europe due to the migration of invasive species.

There is also a dramatic increase in neophytes in the tropics, but the number of small-scale, displaced endemics (which are almost completely absent in Germany, for example) is incomparably higher there. Not only in terms of plant species, Europe from the Atlantic to the Urals already has a much higher biodiversity in the Anthropocene than ever before in the history of the Holocene—while tropical India or Hawaii have a high loss of species (Barthlott 2025).

This means, however, that the results of scientific research on climate change now seem to be widely accepted not only by the scientific community, but also by the general public: "*Since the AR3, progress in understanding, how climate is changing in space and time, has been achieved through improvements and extensions of many data sets and data analyses, broader geographical coverage, better understanding of uncertainties, and a greater variety of measurements*" (IPCC 2007). The AR4 emphasised that "*warming of the climate system is unequivocal, as is now evident from observations of increases in global average air and ocean temperatures, widespread melting of snow and ice, and global average sea level rise*" (IPCC 2007, 2023).

Current knowledge of climate change leads to best estimates and ranges of likelihood for global surface temperature warming for the six IPCC SRES scenarios. All global and regional simulations under scenarios with increasing greenhouse gas concentrations and sulphate aerosols indicate significant increases in both precipitation and temperature towards the end of the twenty-first century. However, the magnitude and regional pattern of these increases are more variable globally. For 2090, the best estimate of the low-warming **IPCC SRES B1** scenario predicts a surface warming of 1.8 K compared to 2000, with a probability range of 1.1–2.9 K.

Table 6.1 shows the global warming for different IPCC SRES scenarios and their corresponding likelihood ranges compared to the constant year 2000 climate.

There is a range of models (Table 6.2) that estimate the magnitude of climate change under different CO_2 and sulphate emissions. However, it must be stressed that the large differences between models on a global scale and the resulting uncertainties must be taken into account.

With regard to the "Climate-Soil-Vegetation System" (Walter 1970; Breckle 2002; Lauer and Rafiqpoor 2002; Breckle and Rafiqpoor 2022), climate is the main factor that determines, in particular, the large-scale distribution of natural vegetation

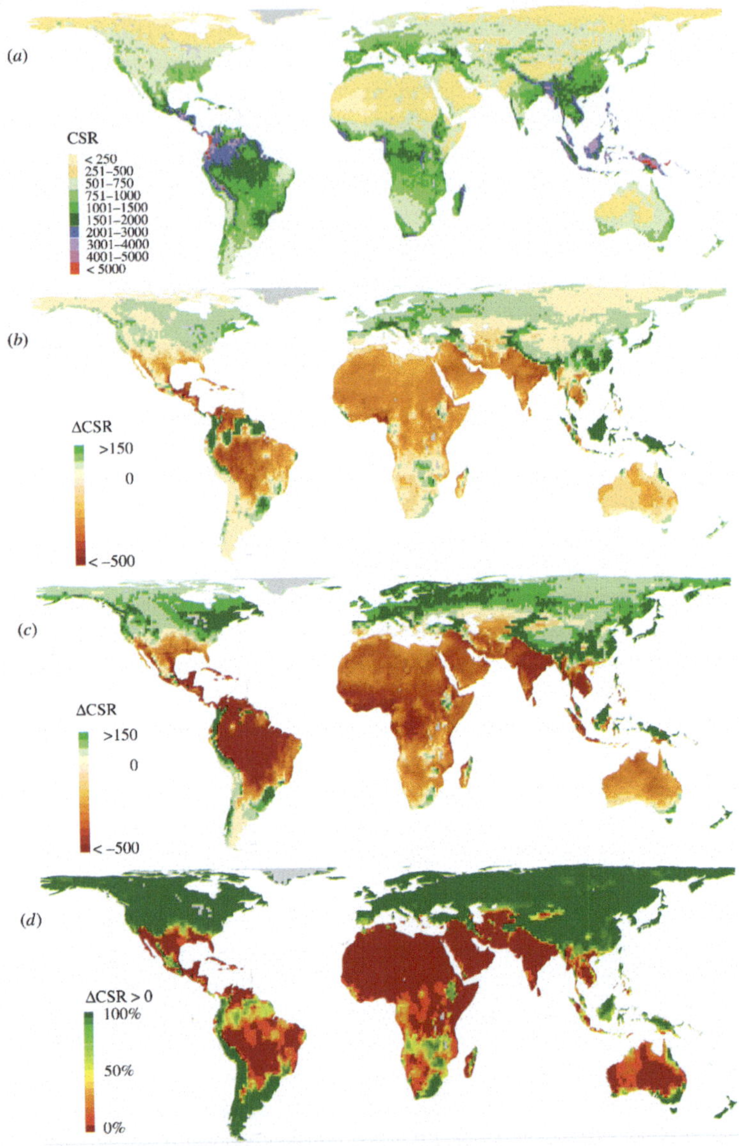

Fig. 6.3 Modelled current global patterns of the capacity for species richness (CSR; species number per $110 \times 110\,\text{km}^2$) and future changes. **a** Modelled current patterns of CSR, **b** change in CSR under + 1.8 K/B1 scenario for 2100, and **c** change in CSR under + 4.0 K/A1FI scenario for 2100. CSR changes are counted in species numbers per $110 \times 110\,\text{km}^2$ grid cell and represent mean values for the CGCM2, CSIRO2, HadCM3 and PCM GCMs. Colour classes represent steps of 50 species. **d** Congruence in the direction of change (either increase in CSR or decrease in CSR, independent from the magnitude of change) between present and future CSR for all 18 available combinations of five GCMs (CGCM2, CSIRO2, ECHAM4 A2/B2, HadCM3 and PCM) and the four major IPCC scenarios (A1FI, A1, B1, B2). The dark green colour stands for 100% congruence across all 18 models that CSR is going to increase, whereas dark red indicates 100% congruence across models that CSR will decrease in the respective area. Yellow areas are subject to oppositional predictions of the direction of change across the models (from Sommer et al. 2010)

Table 6.1 Global average surface warming at the end of the twenty-first century (for further descriptions: see IPCC 2007)

IPCC Scenario	Temperature Change (K at 2090–2099 relative to 1980–1999)	
	Best estimate	Likelihood range
Constant Year 2000	**0.6**	**0.3–0.9**
B1	**1.8**	**1.1–2.9**
A1T	2.4	1.4–3.8
B2	2.4	1.4–3.8
A1B	2.8	1.7–4.4
A2	**3.4**	**2.0–5.4**
A1F1	4.0	2.4–6.4

physiognomy and species composition (Schmithüsen 1976; De Castro et al. 2007). Although it is impossible to precisely predict the potential impacts of climate change on ecosystems in the twenty-first century, the use of selected scenarios representative of possible climate changes as a basis for calculating an ecoclimate classification (see Part I) can provide answers to the question "**how large will the shift of climatic zones be with respect to the changes in global temperature and precipitation predicted by different models for the two selected time slices of the twenty-first century**".

Recent studies on the impacts of climate change on biodiversity indicate that a combination of ecological modelling, i.e. statistical "niche-based" models, with different scenarios of climate change is crucial for determining the environmental conditions that currently drive species distributions (Thuiller et al. 2005; Thuiller 2007). A large number of studies on "Climatic Envelope" modelling, which describes the climate where a species currently lives (its climate "envelope", formerly described as the "ecological amplitude"), and then map the geographic shift of that envelope under climate change, suggest that global climate change is the main driver for the shift in the geographical range of species as a consequence of shifts in normal temperature and precipitation patterns, which generally delimit species boundaries under the competitive conditions between species within vegetation formations (Hamann and Wang 2006; Hamann and Aitken 2012; Thuiller 2007).

Changes in the coupled land–atmosphere system associated with patterns that characterise the states of both vegetation and the atmosphere must be included in analyses of climate change (Fraedrich et al. 2001). When considering the effects of climate change on vegetation, it is widely accepted in the scientific community that shifts in vegetation zones are not entirely determined by shifts in climate. Topography, soil texture, soil moisture, competition between species, and other factors can also influence the distribution of plant species (Talkkari and Hypén 1996; Iverson and Prasad 1998, 2001; Mathys et al. 2014). On the other hand, it is impossible, and apparently never necessary, to account for all of these factors when modelling the effects of climate change on large geographic scales. The abiotic factors (geodiversity) (Lauer and Rafiqpoor 2001), such as climate, can be considered to operate

Table 6.2 Models and abbreviations of various GCMs currently in use (http://climate.calcommons.org/list-models)

[GCMs considered in this study]

Model abbreviation	Model developing research institution
bccr_bcm2_0	Bjerknes Centre for Climate Research, Norway, BCM2.0 Model
cccma_cgcm3_1	Canadian Centre for Climate Modelling and Analysis, CGCM3.1 Model, T47 resolution
cccma_cgcm3_1_t63	Canadian Centre for Climate Modelling and Analysis, CGCM3.1 Model, T63 resolution
miroc3_2_hires	CCSR/NIES/FRCGC, Japan, MIROC3.2, high resolution
miroc3_2_medres	CCSR/NIES/FRCGC, Japan, MIROC3.2, medium resolution
csiro_mk3_0	CSIRO Atmospheric Research, Australia, Mk3.0 Model
csiro_mk3_5	**CSIRO Atmospheric Research, Australia, Mk3.5 Model**
ukmo_hadcm3	Hadley Centre for Climate Prediction, Met Office, UK, HadCM3 Model
ukmo_hadgem1	**Hadley Centre for Climate Prediction, Met Office, UK, HadGEM1 Model**
ingv_echam4	INGV, National Institute of Geophysics and Volcanology, Italy, ECHAM 4.6 Model
inmcm3_0	Institute for Numerical Mathematics, Russia, INMCM3.0 Model
ipsl_cm4	IPSL/LMD/LSCE, France, CM4 V1 Model
iap_fgoals1_0_g	LASG, Institute of Atmospheric Physics, China, FGOALS1.0_g Model
mpi_echam5	**Max Planck Institute for Meteorology, Germany, ECHAM5 / MPI OM**
cnrm_cm3	Meteo-France, Centre National de Recherches Meteorologiques, CM3 Model
miub_echo_g	Meteorological Institute of the University of Bonn, ECHO-G Model
mri_cgcm2_3_2a	Meteorological Research Institute, Japan, CGCM2.3.2a
giss_aom	NASA Goddard Institute for Space Studies, C4 × 3
giss_model_e_h	NASA Goddard Institute for Space Studies, ModelE20/HYCOM
giss_model_e_r	NASA Goddard Institute for Space Studies, ModelE20/Russell
ncar_ccsm3_0	National Centre for Atmospheric Research, CCSM3.0
ncar_pcm1	National Centre for Atmospheric Research, PCM1
gfdl_cm2_0	NOAA Geophysical Fluid Dynamics Laboratory, CM2.0 Model
gfdl_cm2_1	NOAA Geophysical Fluid Dynamics Laboratory, CM2.1 Model

at a higher level in the hierarchy than the biotic factors. "*They have been usefully applied to identify possible magnitudes of future distribution changes and to suggest which species, habitats and regions are most at risk from climate change*" (Pearson and Dawson 2003: 369). Ecosystem-based "climate envelope" modelling approaches have been used to assess potential impacts of climate change on forest communities and tree species (Hamann and Wang 2006; Hamann and Aitken 2012).

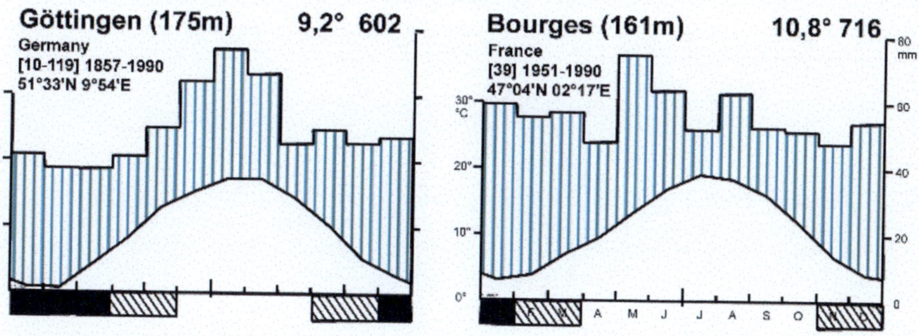

Fig. 6.4 "Twin climatic diagrams" of Göttingen (Germany) and Bourges (France) which exhibit the possible future changes in temperature and precipitation

More effective than the "climate envelope models" (Pearson and Dawson 2003; Hamann and Wang 2006; Hamann and Aitken 2012), however, is the ecoclimate classification (see Part I) as a synthetic approach. The ecoclimate classification is based on the interactions between climate and vegetation and therefore takes into account the relevant geographic factors and ecological requirements of plants via potential evapotranspiration (PET). The classification can help to accurately quantify the potential distribution of vegetation cover under future climate change IPCC SRES scenarios through the analysis of climate zone shifts.

Thus, a simple approach is the comparison of "twin climatic diagrams of Göttingen", meaning diagrams which exhibit the future changes in temperature and precipitation. As an example, Göttingen (Germany), according to our analysis (compare Fig. 6.4), will have within few decades a future climate which is exhibited today by the present diagram of Bourges in France (Fig. 6.4, right) about 735 km to the south-west (Fig. 6.4). Those extrapolations are a first approach for future scenarios (Breckle 2005).

The annual data of long-term observations give a very good picture on the fluctuation of the climate parameters and probably shifting gradients, this can be visualized by 'Climatograms'. They indicate very good the annual oscillations and enable simple calculation of statistic trends. This is exemplified here by the data from the Solling climatic station (Fig. 6.5).

The climate in Solling (Natural Park in Lower Saxony, Germany), which is a typical beech forest vegetation climate, is characterised by a high degree of oceanity with a moderate temperature amplitude between seasons. The long-term observation trend demonstrated by the yearly climatograms 1960–2024 (Fig. 6.7) indicate some significant but not strong differences between individual years. Arid months are seldom, hyperhumid months are common in each year but with no regular maxima. Such long-term data enable us to see trends. The increase in the mean annual temperature (see Fig. 6.5) from 1960 to 2024 is already distinctly more than 1.5 K, the trend in mean annual precipitation (see Fig. 6.6) with a slight increase is less clear.

These days a huge amount of data are available, especially interpolated and simulated climate data for raster cells from most parts of the globe with long temporal

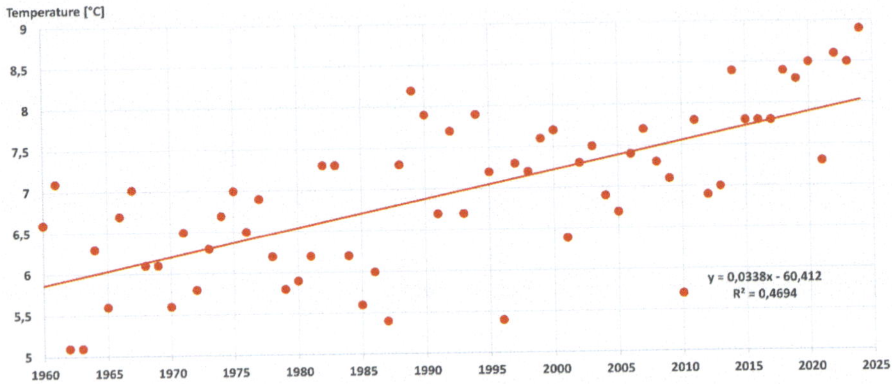

Fig. 6.5 Increase in the mean annual temperature in Solling (Germany) from 1960 to 2024

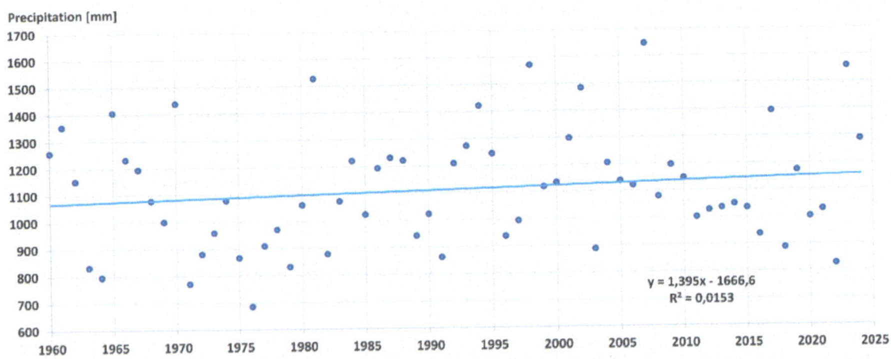

Fig. 6.6 Mean annual precipitation in Solling (Germany) for the period 1960 to 2024

coverages (often starting in 1900) and spatial resolution of 0.5° × 0.5° or even higher (Harriss et al. 2023).

However, an important basis for all the meteorological and climatic data are the real measured weather data from weather stations around the world without rounding up and downs and interpolations. Figure 6.7 is an excellent example from one station in Central Germany. Thousands of stations support worldwide monitoring of current climatic conditions.

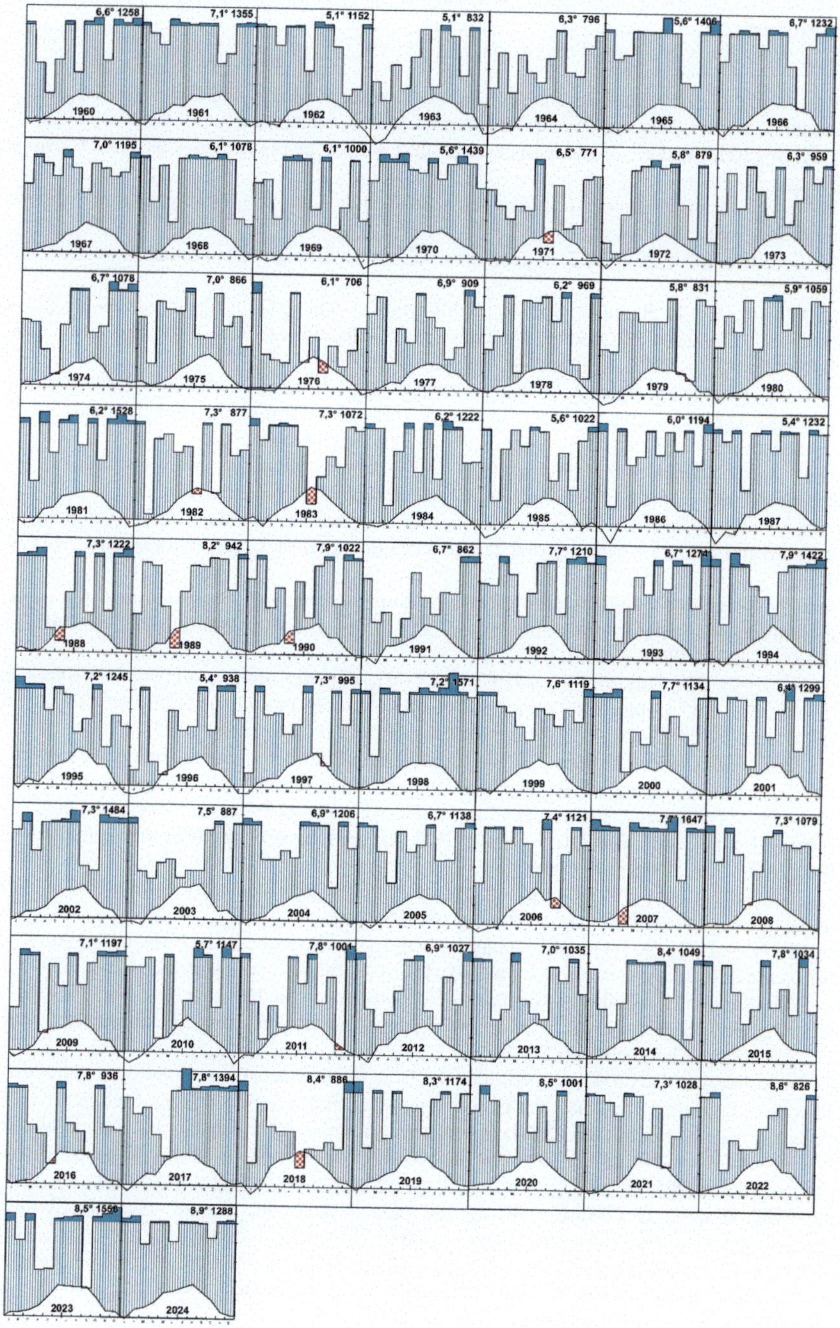

Fig. 6.7 The year-to-year variation in temperature and precipitation in Solling (Germany) for the period 1960 to 2024 shown using ecological climatograms

References

Barthlott W (2025) Abschlussbericht der Kommission für Biologie und Geowissenschaften der Akademie der Wissenschaften und der Literatur Mainz, Stuttgart, p 5–20

Breckle S-W (2002) Walter's Vegetation of the earth. The ecological systems of the Geo–Biosphere. Springer, Berlin

Breckle S-W (2005) Möglicher Einfluss des Klimawandels auf die Waldvegetation Nordwest-deutschlands? LÖBF-Mitteilungen 2(05):12–17

Breckle S-W, Rafiqpoor MD (2022) Vegetation and Climate. Springer, Heidelberg, New York. ISBN 789-3-662-64035-7

Crippa M, Guizzardi D, Banja M, Solazzo E, Muntean M et al (2022) CO_2 emissions of all world countries. JCR/IEA/PBL Report. Luxembourg: Publications office of the European Union. ISBN: 978-92-76-55802-6. https://doi.org/10.2760/07904

De Castro M, Gallardo C, Jylha K, Tuomenvirta H (2007) The use of a climate-type classification for assessing climate change effects in Europe from an ensemble of regional climate models. Clim Change 81:329–341. https://doi.org/10.1007/s10584-006-9224-1

Fraedrich K, Gerstengarbe FW, Werner PC (2001) Climate shifts during the last century. Clim Change 50:405–417. https://doi.org/10.1023/A:1010699428863

Hamann A, Aitken SN (2012) Conservation planning under climate change: accounting for adaptive potential and migration capacity in species distribution models. Biodiversity Distrib 19(3):268–280

Hamann A, Wang T (2006) Potential effects of climate change on ecosystem and tree species distribution in British Columbia. Ecology 87(11):2773–2786. https://doi.org/10.1890/0012-9658(2006)87[2773:PEOCCO]2.0.CO;2

Harris IC, Jones PD, Osborn T (2023) CRU TS4.07: climatic research unit (CRU) time-series (TS) version 4.07 of high-resolution gridded data of month-by-month variation in climate (Jan. 1901–Dec. 2022). NERC EDS Centre for Environmental Data Analysis, 22.02.2024

IPCC (2007) AR4 Climate change 2007—synthesis report. https://www.ipcc.ch/report/ar4/syr/

IPCC (2023) Climate change 2023: synthesis report. Contribution of Working Groups I, II and III to the Sixth Assessment Report of the Intergovernmental Panel on Climate Change [Core Writing Team, Lee H, Romero, J. (eds)]. IPCC, Geneva, Switzerland https://doi.org/10.59327/IPCC/AR6-9789291691647.001

Iverson LR, Prasad, AM (1998) Predicting abundance of 80 tree species following climate change in the eastern United States. Ecol Monogr 68:465–485

Iverson L, Prasad AM (2001) Potential changes in tree species richness and forest community types following climate change. Ecosystems 4(3):186–199. https://www.jstor.org/stable/3658952

Lauer W, Rafiqpoor MD (2001) Biodiversität - Geodiversität - Ökodiversität. Die Artenvielfalt – der lebendige Reichtum der Erde. In: Meyers Großer Weltatlas. Atlas zum 21. Jahrhundert. Meyers Lexikonverlag, Mannheim, pp 28–29

Lauer W, Rafiqpoor MD (2002) Die Klimate der Erde – eine Klassifikation auf der Grundlage der ökophysiologischen Merkmale der realen Vegetation. Franz Steiner Verlag, Stuttgart, p 271

Mathys A, Coops N, Waring RH (2014) Soil water availability effects on the distribution of 20 tree species in western North America. For Ecol Manag 313(4):144–152. https://doi.org/10.1016/j.foreco.2013.11.005

Pearson RG, Dawson TP (2003) Predicting the impacts of climate change on the distribution of species: are bioclimate envelope models useful? Global Ecol Biogeogr 12:361–371. https://www.researchgate.net/publication/227549255_Predicting_the_impacts_of_climate_change_on_the_distribution_of_species_Are_bioclimate_envelope_models_useful

Schmitthüsen J (1976) Atlas zur Biogeographie. Mannheim

Sommer JH, Kreft H, Kier G, Jetz W, Mutke J, Barthlott, W (2010) Projected impacts of climate change on regional capacities for global plant species richness. Proc R Soc A B 277:2271–2280. https://doi.org/10.1098/rspb.2010.0120. http://rspb.royalsocietypublishing.org/content/early/2010/03/20/rspb.2010.0120

Talkkari A, Hypén H (1996) Development and assessment of a gap–type model to predict the effects of climate change on forests based on spatial forest data. For Ecol Manage 83(3):217–228. https://doi.org/10.1016/0378-1127(96)03719-X

Thuiller W (2007) Biodiversity: climate change and the ecologist. Nature 448(7153):550–552. https://doi.org/10.1038/448550a. PMID: 17671497

Thuiller W, Lavorel S, Araújo MB (2005) Niche properties and geographical extent as predictors of species sensitivity to climate change. Glob Ecol Biogeogr 14:347–357

Wiegandt K (ed) (2022) 3 Grad mehr. Ein Blick in die drohende Heißzeit und wie uns die Natur helfen kann, sie zu verhindern. Oekom, München

Chapter 7
The Objectives of the Ecoclimate Classification Model

Although there is an abundance of literature (see e.g., IPCC 2007, 2023) on the causes and effects of global changes in temperature and precipitation as a result of climate change (Houghton 2001; Williams et al. 2007; Woodward et al. 2004; Leemans et al. 1996; Parmesan & and Yohe 2003; Kelly and Goulden 2008), there are no or only a few comprehensive studies that analyse and assess **globally landscape changes** resulting from the combination of changes in heat and water budgets caused by climate change. In this context, only individual small-scale studies based mainly on "climate envelope" modelling are available to assess the potential impacts of climate change for Europe (i.e., Van Minnen et al. 2002; Vermaat et al. 2016; Räisänen et al. 2004; Thuiller et al. 2005; Seneviratne et al. 2006; De Castro et al. 2007; Illés and Móricz 2022), for North America (Iverson and Prasad 1998; Lugo et al. 1999; Iverson and Prasad 2001; Hamann and Wang 2006; Cayan et al. 2008; Hamann & and Aitken 2012; McFarland Bell et al. 2014; Balzotti et al. 2016), for the polar regions (Hinzman 2005; Day et. al 2024; Nelson et al. 2002; Roderfeld et al. 2008, and the cited literature), for the Mediterranean region and the Middle East (Giorgi and Lionello 2008; Evans et al. 2004; Evans 2009; Planton 2012), or also for India (Murari et al. 2001; Ravindranath et al. 2006; Rupa Kumar et al. 2006; Dobler and Ahrens 2011), and China (Liu and Chen 2000; Bueh 2003; Cholaw 2003; Weng et al. 2006; Guo and Wang 2011; Liu et al. 2013; Li et al. 2020), etc. The efforts of Fraedrich et al. (2001), Kalvova et al. (2003), Beck et al. (2005), Beck et al. (2023), Kottek et al. (2006), etc. to simulate changes in climatic regions in the course of climate change on the basis of the climate classification of (Köppen 1900) have been discussed in detail in Part I.

In this section (Part II), the method of ecoclimate classification (see Part I) is applied to some climate change SERS scenarios of IPCC and the spatial changes in ecoclimates resulting from the changed temperature, precipitation and ETP data for the middle (2050) and end (2090) of the twenty-first century. It will be shown that the ecoclimate classification is a very useful tool for estimating landscape changes (i.e., shift of climate zones) due to climate change, as it is based on the household

M. D. Rafiqpoor and S.-W. Breckle, *The World's Ecoclimates*,
SpringerBriefs in Earth System Sciences,
https://doi.org/10.1007/978-3-032-06456-1_7

(or balance) variables of the hygrothermal climate in the "Climate-Soil-Vegetation System" as a prerequisite for the growth of the Earth's natural plant cover (Lauer and Rafiqpoor 2002).

References

Balzotti CS, Kitchen SG, McCarthy C (2016) Beyond the single species climate envelope: a multifaceted approach to mapping climate change vulnerability. Ecosphere 7(9):e01444. 23 pp. https://doi.org/10.1002/ecs2.1444

Beck C, Grieser J, Kottek M, Rubel F, Rudolf B (2005) Characterizing global climate change by means of Köppen climate classification. In: Klimastatusbericht: 139–149. Deutscher Wetterdienst, Offenbach. https://www.researchgate.net/publication/255532038

Beck HE, McVicar TR, Vergopolan N, Berg A, Lutsko NJ et al (2023) High-resolution (1 km) Köppen-Geiger maps for 1901–2099 based on constrained CMIP6 projections. Scientific Data 10, Article number: 724, pp 16. https://doi.org/10.1038/s41597-023-02549-6

Bueh C (2003) Simulation of the future change of east Asian monsoon climate using the IPCC series A2 and B2 scenarios. Chin Sci Bull 48(10):1024–1030. https://doi.org/10.1007/BF03184220

Cayan DR, Maurer EP, Dettinger MD, Tyree M, Hayhoe K (2008) Climate change scenarios for the California region. Climatic Change, 87. Suppl 1:21–42. https://doi.org/10.1007/s10584-007-9377-6

Day JJ, Svensson G, Casati B, Uttal T, Khalsa S-J et al (2024) The Year of Polar Prediction site Model Intercomparison Project (YOPPsiteMIP) phase 1: project overview and Arctic winter forecast evaluation. Geoscientific Model Dev 17:5511–5543. https://doi.org/10.5194/gmd-17-5511-2024

De Castro M, Gallardo C, Jylha K, Tuomenvirta H (2007) The use of a climate-type classification for assessing climate change effects in Europe from an ensemble of regional climate models. Clim Change 81:329–341. https://doi.org/10.1007/s10584-006-9224-1

Dobler A, Ahrens B (2011) Four climate change scenarios for the Indian summer monsoon by the regional climate model COSMO–CLM. J Geophys Res Atm 116(D24):24104: 1–13. https://doi.org/10.1029/2011JD016329

Evans JP (2009) 21st century climate change in the Middle East. Clim Change 92:417–432. https://doi.org/10.1007/s10584-008-9438-5

Evans JP, Smith RB, Oglesby RJ (2004) Middle East climate simulation and dominant precipitation processes. Int J Climatol 24:1671–1694. https://doi.org/10.1002/joc.1084

Fraedrich K, Gerstengarbe FW, Werner PC (2001) Climate Shifts during the last century. Clim Change 50:405–417. https://doi.org/10.1023/A:1010699428863

Giorgi F, Lionello P (2008) Climate change projections for the mediterranean region. Global Planet Change 63:90–104. https://doi.org/10.1016/j.gloplacha.2007.09.005

Guo D, Wang H (2011) The significant climate warming in the northern Tibetan plateau and its possible causes. Intern J of Climatol 32(12):1775–1781. https://doi.org/10.1002/joc.2388

Hamann A, Aitken SN (2012) Conservation planning under climate change: accounting for adaptive potential and migration capacity in species distribution models. Biodiversity Distrib 19(3):268–280

Hamann A, Wang T (2006) Potential effects of climate change on ecosystem and tree species distribution in British Columbia. Ecology 87(11):2773–2786. https://doi.org/10.1890/0012-9658(2006)87[2773:PEOCCO]2.0.CO;2

Hinzman LD, Bettez ND, Bolton WR, Chapin FS, Dyurgerov MB et al (2005) Evidence and implications of recent climate change in northern Alaska and other arctic regions. Clim Change 72(3):251–298

Houghton JTEA, Griggs DJ, Linden PJ van der, Noguer M et al (2001) The scientific basis: contribution of working group I to the third assessment report of the intergovernmental panel on climate change. Climate change: 525–582. Cambridge University Press, Cambridge. https://www.researchgate.net/publication/216811760_Climate_Change_2001_The_Scientific_Basis

Illés G, Móricz N (2022) Climate envelope analyses suggest significant rearrangements in the distribution ranges of Central European tree species. Ann for Sci 79(35):19. https://doi.org/10.1186/s13595-022-01154-8

IPCC (2007) AR4 climate change 2007—synthesis report. https://www.ipcc.ch/report/ar4/syr/

IPCC (2023) Climate Change 2023: Synthesis Report. Contribution of Working Groups I, II and III to the sixth assessment report of the intergovernmental panel on climate change [Core Writing Team, Lee H, Romero, J. (eds)]. IPCC, Geneva, Switzerland https://doi.org/10.59327/IPCC/AR6-9789291691647.001

Iverson L, Prasad A (1998) Estimating regional plant biodiversity with GIS modelling. Diversity Distrib 4(2):49–61. https://doi.org/10.1046/j.1472-4642.1998.00007.x

Iverson L, Prasad AM (2001) Potential changes in tree species richness and forest community types following climate change. Ecosystems 4(3):186–199. http://www.jstor.org/stable/3658952

Kalvova J, Halenka T, Bezpalcova K, Nemesova I (2003) Köppen climate types in observed and simulated climates. Stud Geophys Geod 47:185–202

Kelly AE, Goulden ML (2008) Rapid shifts in plant distribution with recent climate change. PNAS 105:11823–11826

Köppen W (1900) Versuch einer Klassifikation der Klimate, vorzugsweise nach ihren Beziehungen zur Pflanzenwelt. Geographische Zeitschrift, vol 6: 593–611 und 657–679

Kottek M, Grieser M, Beck Ch, Rudolf B, Rubel F (2006) World map of the Köppen-Geiger climate classification updated. Meteorologische Zeitschrift 15:259–263

Lauer W, Rafiqpoor MD (2002) Die Klimate der Erde – eine Klassifikation auf der Grundlage der ökophysiologischen Merkmale der realen Vegetation. Franz Steiner Verlag, Stuttgart, p 271

Leemans R, Crame W, Minnen JGV (1996) Prediction of global biome distribution using bioclimatic equilibrium models. In: Breymeyer, Hall DO, Melillo JM, Ågren GI (eds) Global change: effects on coniferous forests and grasslands. John Wiley & Sons Ltd, New York

Li G, Huang J, Guo H, Du S (2020) Projecting species loss and turnover under climate change for 111 Chinese tree species. Forest Ecol Manag 477(1). https://doi.org/10.1016/j.foreco.2020.118488

Liu X, Chen B (2000) Climatic warming in the Tibetan plateau during recent decades. Intern J Climatol 20:1729–1742. https://doi.org/10.1002/1097-0088(20001130)20:14%3c1729::AID-JOC556%3e3.0.CO;2-Y

Liu C, White M, Newell G, Pearson R (2013) Selecting thresholds for the prediction of species occurrence with presence-only data. J Biogeogr 40(4):778–789. https://www.researchgate.net/publication/260161376_Selecting_thresholds_for_the_prediction_of_species_occurrence_with_presence-only_data

Nelson FE, Anisimov OA, Shiklomanov NI (2002) Climate change and hazard zonation in the circum-arctic permafrost regions. Nat Hazards 26:203–225

Parmesan C, Yohe G (2003) A globally coherent fingerprint of climate change impacts across natural systems. Nature 421(6918):37–42. https://doi.org/10.1038/nature01286

Räisänen J, Hansson U, Ullerstig A, Döscher R, Graham LP et al (2004) European climate in the late twenty-first century: regional simulations with two driving global models and two forcing scenarios. Clim Dyn 22:13–31

Ravindranath NH, Joshi NV, Sukumar R, Saxena A (2006) Impact of climate change on forest in India. Curr Sci 90:354–362

Roderfeld H, Blyth E, Dankers R, Huse G, Slagstad D et al (2008) Potential impact of climate change on ecosystems of the Barents Sea region. Clim Change 87:283–303

Rupakumar K, Sahai AK, Krishna Kumar K, Patwardhan SK, Mishra et al (2006) High-resolution climate change scenarios for India for the 21st century. Curr Sci 90:334–345. https://www.researchgate.net/publication/363336598_High-resolution_climate_change_scenarios_for_India_for_the_21st_century
Seneviratne SI, Lüthi D, Litschi M, Schär C (2006) Land-atmosphere coupling and climate change in Europe. Nature 443:205–209
Thuiller W, Lavorel S, Araújo MB, Sykes MT, Prentice IC (2005) Climate change threats to plant diversity in Europe. PNAS 102:8245–8250
Van Minnen JG, Onigkeit J, Alcamo J (2002) Critical climate change as an approach to assess climate change impacts in Europe: Development and application. Environ Sci Policy 5:335–347. https://doi.org/10.1016/S1462-9011(02)00044-8
Vermaat JE, Hellmann FA. van Teeffelen A, van Minnen J, et al. (2016) Differentiating the effects of climate and land use change on European biodiversity: A scenario analysis. AMBIO – A J Human Environ 46(3):14 pp. https://doi.org/10.1007/s13280-016-0840-3
Weng E, Zhoua G (2006) Modelling distribution changes of vegetation in China under future climate change. Environ Model Assess 11:45–58. https://doi.org/10.1007/s10666-005-9019-1
Williams JW, Jackson ST (2007) Novel climates, no-analog communities, and ecological surprises. Front Ecol Environ 5(9):475–482. https://doi.org/10.1890/070037
Woodward FI, Lomas MR (2004) Vegetation dynamics: simulating responses to climatic change. Biol Rev Camb Phil Soc 79:643–670. https://doi.org/10.1017/s1464793103006419

Chapter 8
Selection of Scenarios and Models

The pattern of today's ecoclimate types on Earth is shown in Fig. 4.8 (including the corresponding legend in Fig. 4.9). Here three **GCMs** were selected to demonstrate the spatial shift of ecoclimatic zones on Earth due to climate change and their variations within the climate models (see Table 6.2, marked in **bold**):

- **CSIRO_MK3_5 Model**: Institute for Atmospheric Research, Australia
- **UKMO_HADCM3 Model**: Hadley Centre for Climate Prediction, Meteor. Office, UK
- **MPI_ECHAM5 Model**: Max-Planck Institute for Meteorology, Germany

In order to determine the extent of impact of climate change on global climate zone shifts, from Table 6.1 scenarios were selected for the **smallest** and **largest** temperature and precipitation changes. The **IPCC SRES Scenarios B1** and **A2** for the time slices 2050 and 2090 were therefore selected. The **IPCC SRES A1F1** scenario with the highest possible temperature increase around 2100 unfortunately had to be eliminated, as Conservation International (https://www.worldclim.org/data/cmip6/cmip6climate.html) does not provide any data for this scenario.

The **B1 storyline** and scenario family describes a convergent world with the same global population, peaking in the middle of the century and declining thereafter, but with a rapid transformation of economic structures towards a service and information economy, with a reduction in material use intensity and the introduction of clean and resource-efficient technologies. The focus is on global solutions for economic, social and environmental sustainability, including improved equity, but without additional climate initiatives (IPCC 2007). It might sound somehow utopic, taking the present political situations into account.

The **A2 storyline** and scenario family describes a very heterogeneous world. The underlying theme is self-reliance and the preservation of local identities. Fertility patterns between regions converge very slowly, resulting in a continuously growing population. Economic development is primarily regional, and per capita economic

M. D. Rafiqpoor and S.-W. Breckle, *The World's Ecoclimates*,
SpringerBriefs in Earth System Sciences,
https://doi.org/10.1007/978-3-032-06456-1_8

growth and technological changes are more fragmented and slower than in other storylines (IPCC 2007).

Reference

IPCC (2007) AR4 climate change 2007—synthesis report. https://www.ipcc.ch/report/ar4/syr/

Chapter 9
Material and Method

With the help of the monthly data on temperature, precipitation and potential evapotranspiration, the length of the thermal and hygric vegetation period can be calculated as ecologically relevant variables of the heat and water balance and transferred to an ecoclimate classification system (see Fig. 4.8). The monthly values of temperature and precipitation were taken from the freely available database of Conservation International "for future climatology" for the **IPCC SRES B1** and **A2 scenarios** for the three selected models and the time slices 2050 (mean monthly climatology 2041–2060) and 2090 (mean monthly climatology 2081–2100).

Conservation International, in collaboration with the University of Madison, Wisconsin, the National Centre for Ecological Analysis and Synthesis (NCEAS), and the University of California, Santa Barbara, produced a downscaled and debiased dataset with sufficient spatial resolution of 5 km (2.5 min) based on the CRU (New et al. 2002) and **WorldClim** (Hijmans et al. 2005) datasets and made it available to the international community for further analysis "*for evaluating the potential impacts of twenty-first century climate change on species distributions and ecological processes*". The data available here are **CMIP6** (Coupled Model Intercomparison Project of World Climate Research Programme) downscaled future climate projections. The downscaling and calibration (bias correction) was done with WorldClim v2.1 as baseline climate (Tabor and Williams 2010) (https://www.worldclim.org/data/cmip6/cmip6climate.html).

While the length of the thermal growing season (in months) could be determined by selecting a threshold value from the temperature data, monthly precipitation and potential evapotranspiration (PET) data are required to calculate the length of the hygric growing season as the climatic water balance. Unfortunately, no comparable data on "future climatology" is available for PET in the Conservation International database. Aware that the use of present PET data to calculate future climates underestimates the actual extent of climatic changes and their spatial variation (i.e., shift of climate zones), we have used the PET data of FAO (Zomer and Trabucco 2022) for "current climatology" to calculate the water balance for the climate change scenarios

M. D. Rafiqpoor and S.-W. Breckle, *The World's Ecoclimates*,
SpringerBriefs in Earth System Sciences,
https://doi.org/10.1007/978-3-032-06456-1_9

due to a lack of "future PET data". Despite this shortcoming, the combination of the data sets used results in fascinating images of the global ecoclimatic changes as an expression of the spatial arrangement of ecoclimate types. In ecological climatic diagrams (Walter and Lieth 1960–67) the temperature data are used as an indication for the evaporation data. Of course, this is only a rough, but helpful approximation.

The ecoclimate classification models for the **B1** and **A2** scenarios were calculated for the three **GCMs** mentioned above (see Table 6.2, marked in **bold**). It was found that although all three **GCMs** predict a significant shift in climate zones in the two scenarios for the selected time slices (see Table 6.1), there are considerable differences between the models and the comparable scenarios themselves.

References

Hijmans RJ, Cameron SE, Parra JL, Jones PG, Jarvis A (2005) Very high-resolution interpolated climate surface for global land areas. Intern J Climatol 25:1965–1978. https://doi.org/10.1002/joc.1276

New M, Lister D, Hulme M, Makin I (2002) A high-resolution data set of surface climate over global land areas. Climate Res 21:1–25

Tabor K, Williams JW (2010) Globally downscaled climate projections for assessing the conservation impacts of climate change. Ecol Appl 20(2):554–565

Walter H, Lieth H (1960–67): Klimadiagramm-Weltatlas. VEB Gustav Fischer Verlag Jena

Zomer R, Trabucco A (2022) Global aridity index and potential evapotranspiration climate database v3—algorithm code (Python). https://doi.org/10.6084/m9.figshare.20005589

Chapter 10
Results—Spatial Patterns of the Shift in Ecoclimatic Zones

Figure 10.1a, b show the ecoclimate maps for the **SERS B1** and **SERS A2** scenarios of IPCC for the selected three models of Table 6.2 (green) for the time slice 2090 compared to today. The **B1** scenarios of the three **GCMs** show significant shifts in the climate zones compared to today. However, these shifts are at a slightly lower level compared to that of the **A2** scenarios due to the fact that temperatures have not yet risen excessively by the middle of the twenty-first century. However, the climate maps of all models of the two scenarios **B1** (see Fig. 10.1a) and **A2** (see Fig. 10.1b) show considerable modifications in comparison to each other and to the current pattern of climate types (see Fig. 4.8).

In order to better highlight the future shifts in ecoclimate zones, we only compare and comment on the image of the **A2 scenario** of the **UKMO_HADCM3 model** for the 2090 time slice with the current image of the ecoclimate zones (see Fig. 4.8) as an example. The reader can see the differences between the individual models by comparing the ecoclimate maps for the respective models in Fig. 10.1a, b.

To better illustrate the climate zone shifts, some key areas from the global pattern of ecoclimate regions are selected as examples where the differences between today's and the simulated image of future ecoclimate zones are particularly striking.

But before we get into the discussion about the extent and spatial variation of climate zone shifts, let's take a look at the extremely aesthetic night-time image of our beautiful planet from a satellite perspective (Fig. 10.2). We can easily see that it is no coincidence, for example, that in North America a striking boundary between the much too bright appearing, densely populated east and the somewhat dark appearing, sparsely populated west (with the exception of the large cities on the Pacific coast) of the USA runs almost along the 100th degree of longitude. Curiously, this line coincides with the **climatic dry limit** in North America (Lauer 1952; Jätzold 1961, 1962; Lauer and Rafiqpoor 2002) and separates the macrotherm-subhumid climates of different categories of the east from the macrotherm-semiarid climates of different climate zones of the west of USA. We can almost imagine the consequences for

M. D. Rafiqpoor and S.-W. Breckle, *The World's Ecoclimates*,
SpringerBriefs in Earth System Sciences,
https://doi.org/10.1007/978-3-032-06456-1_10

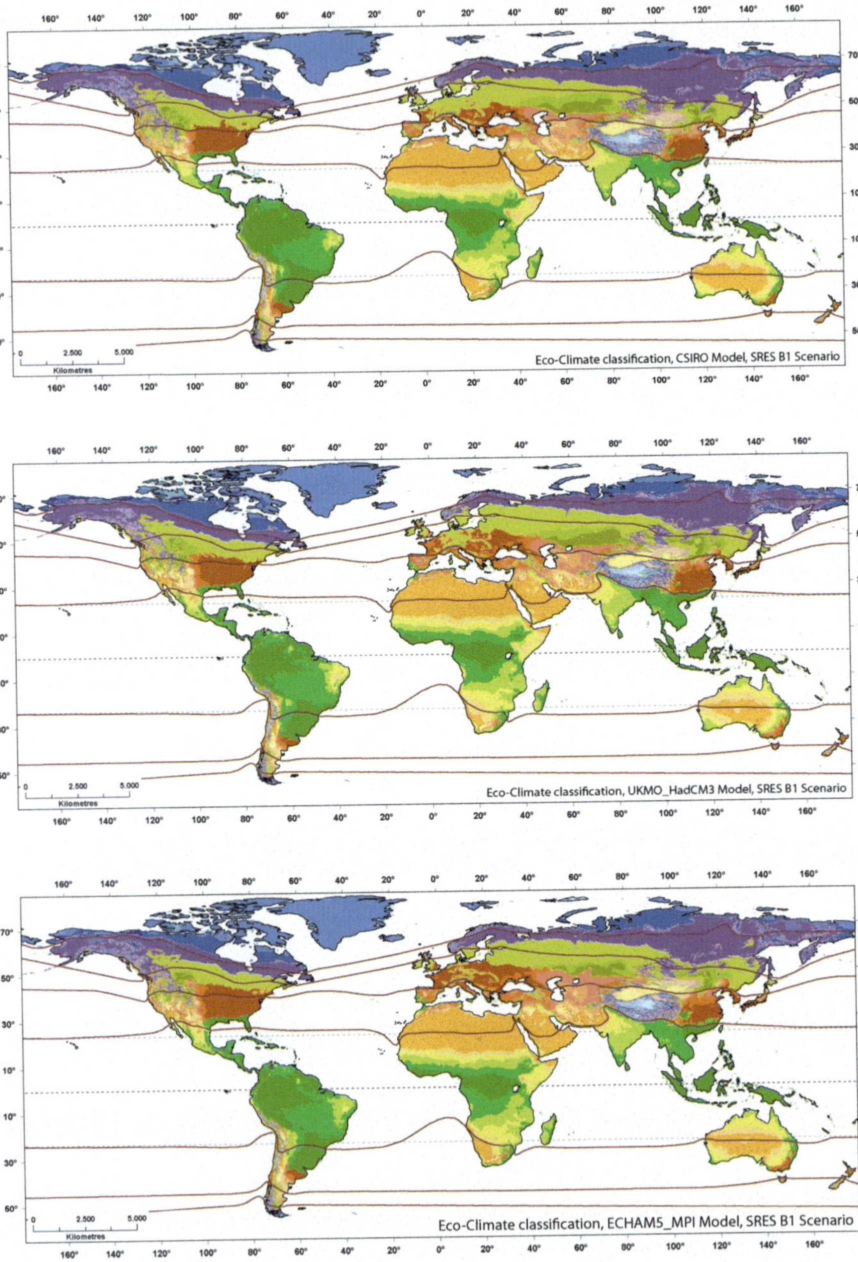

Fig. 10.1 **a** Ecoclimatic classification for **B1 scenarios** of the three selected climate change models (GCMs) for 2090 (for legend and shift of ecoclimatic zones compared to today, see Figs. 4.8 and 4.9)

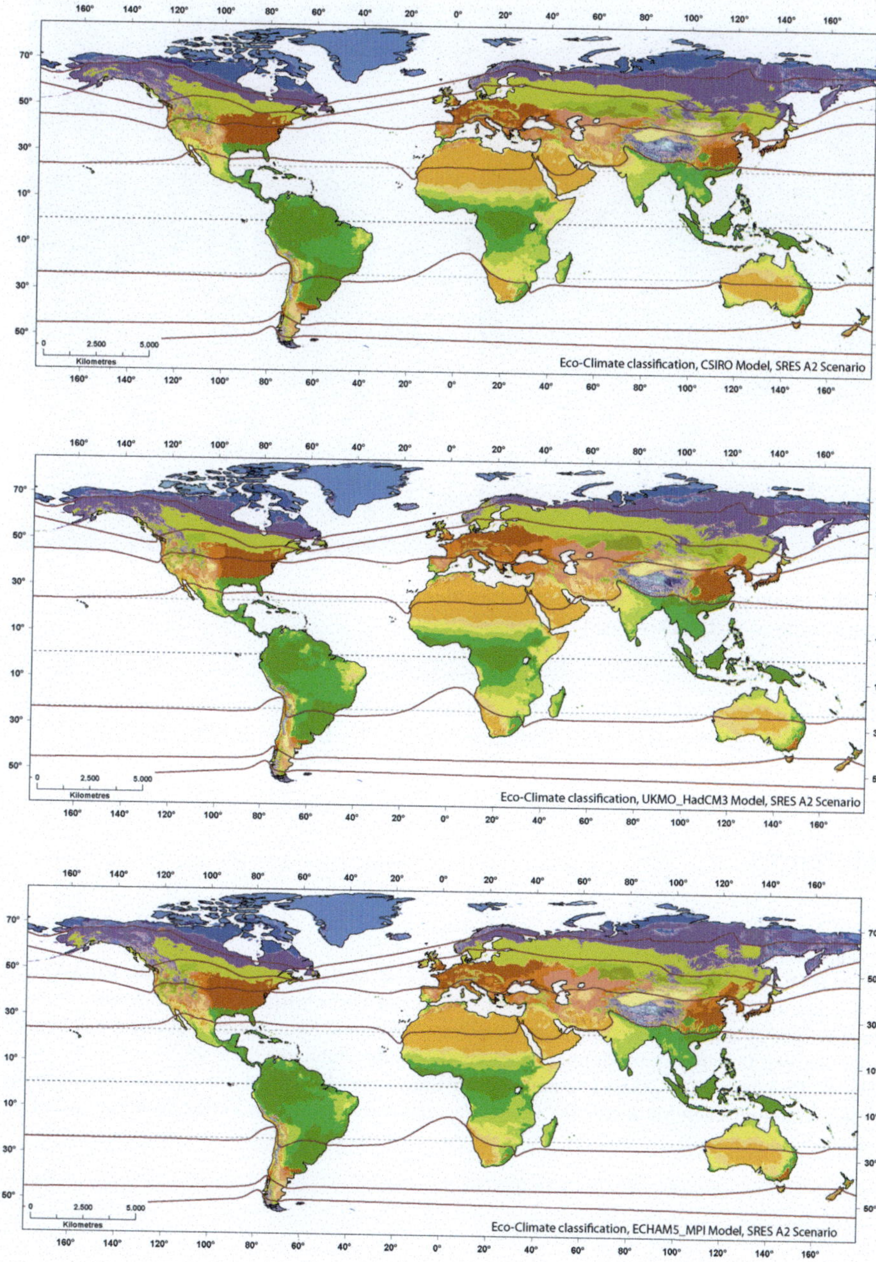

Fig. 10.1 **b** for Ecoclimatic classification for **A2 scenarios** of the three selected climate change models (GCMs) for 2090 (for legend and shift of ecoclimatic zones compared to today, see Figs. 4.8 and 4.9)

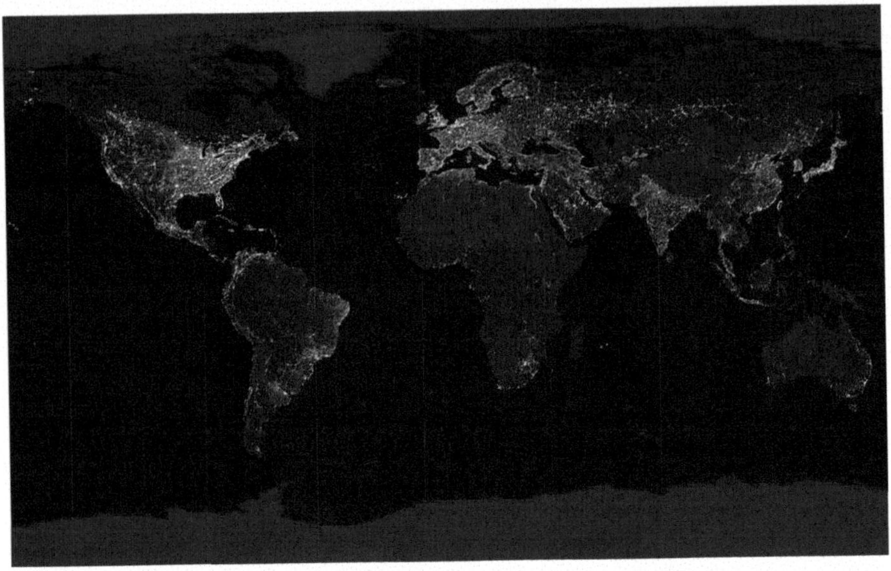

Fig. 10.2 Night image of our planet from a satellite perspective (*Source* https://science.nasa.gov/resource/earth-at-night/; Download November 2024)

humans and nature of the changes in temperature and precipitation regimes (the shift in ecoclimatic zones) along this boundary caused by future climate change.

References

Jätzold R (1961) Aride und humide Jahreszeiten in Nordamerika. Stuttgarter Geographische Studien, Bd. 71, Stuttgart

Jätzold R (1962) Die Dauer der ariden und humiden Zeit des Jahres als Kriterium für Klimaklassifikation. In: Leidlmair A (ed): Hermann v. Wissmann-Festschrift: 89-–08. Tübingen

Lauer W (1952) Humide und aride Jahreszeiten in Afrika und Südamerika und ihre Beziehungen zu den Vegetationsgürteln. Bonner Geographische Abhandlungen 9:15–98

Lauer W, Rafiqpoor MD (2002) Die Klimate der Erde—eine Klassifikation auf der Grundlage der ökophysiologischen Merkmale der realen Vegetation. Franz Steiner Verlag, Stuttgart, p 271

Chapter 11
Regional Aspects of Climate Change—The Shift of Ecoclimate Zones

11.1 North America

It is assumed that a doubling of atmospheric greenhouse gas emission and the resulting changes in the heat and water balance would change the composition of about one third of the planet's forest area (Shriner et al. 1998; Melillo 1999; Melillo et al. 2002, 2011). In the Federal Republic of Germany, for example, it is already apparent that "*climate change has created space for tree species from other biogeographical regions*" (Asche and Stromberg 2024: 59). These changes are even more pronounced on a continental scale. We illustrate this by comparing the current patterns of climate zone arrangement with the **IPCC SERS A2** scenario of the **UKMO_HADCM3** model for 2090.

In **North America** (Fig. 11.1a, b), all climate zones will experience a dramatic northward shift by the end of the twenty-first century (Iverson and Prasad 1998; Iverson and Prasad 2001; Iverson et al. 2008). Primarily the zonoecotones and orobiomes will be subject to considerable change, while the central parts of the zonobiomes will remain relatively stable even with rising temperatures (~4 K by 2100) and changes in precipitation. Megatherm-hyperhumid conditions are expected throughout the fertile eastern USA east of the above-mentioned **climatic dry limit** up to about 45° N. In the southern states (Texas, Louisiana, Mississippi, Alabama, Florida, and even the southern parts of Georgia and South Carolina), tropical conditions with a megatherm-hyperhumid climate may develop.

The effects of climate change are also more serious in the small north-eastern states of the USA (Maine, New Hampshire, Vermont, Rhode Island, Connecticut, New York, Pennsylvania), where megatherm-hyperhumid subtropical climates prevail about 5° of latitude to the north.

This coincides with the findings of other authors who, taking into account the SERS A2 scenario at the end of the twenty-first century, predict an extended growing season of around six additional weeks for this region (Frumhoff et al. 2007). The shift in climate zones and the lengthening of the vegetation period are accompanied by

M. D. Rafiqpoor and S.-W. Breckle, *The World's Ecoclimates*,
SpringerBriefs in Earth System Sciences,
https://doi.org/10.1007/978-3-032-06456-1_11

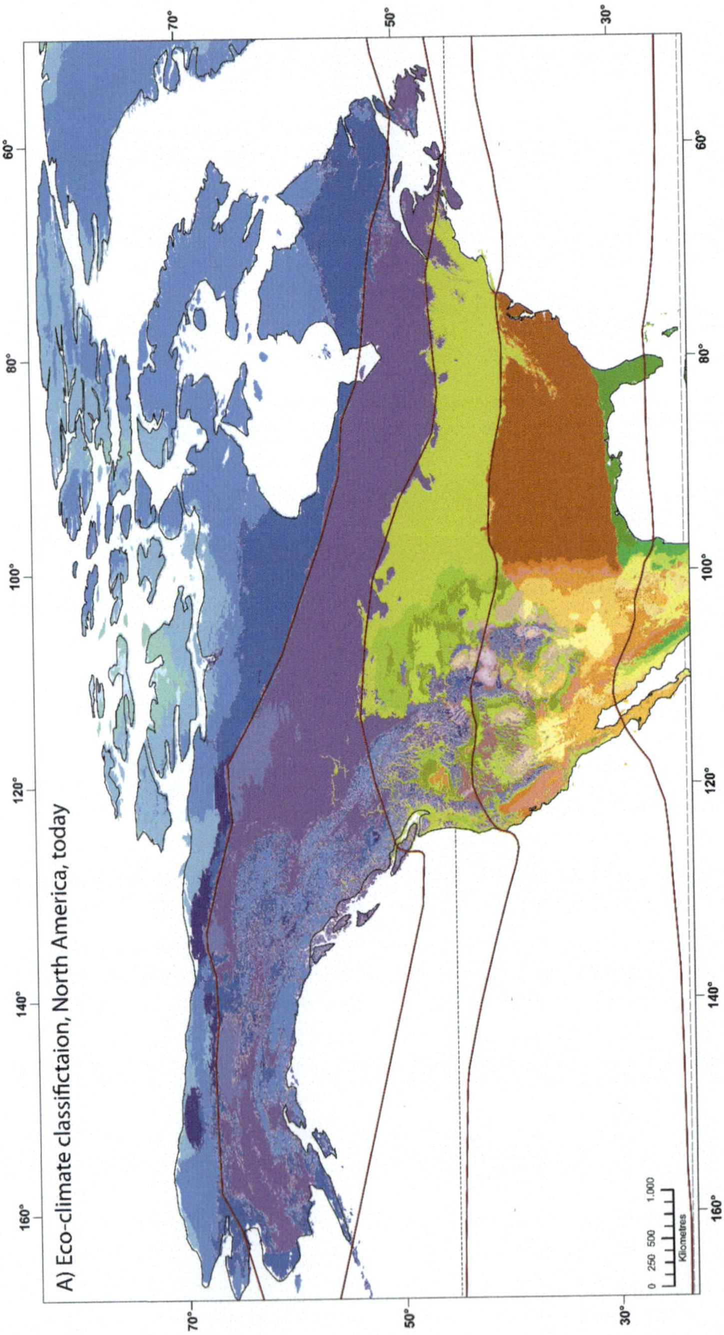

Fig. 11.1 **a** Ecoclimatic zones of North America today. (Lines indicate the boundaries of recent climate zones) (for legend see Figs. 4.8 and 4.9)

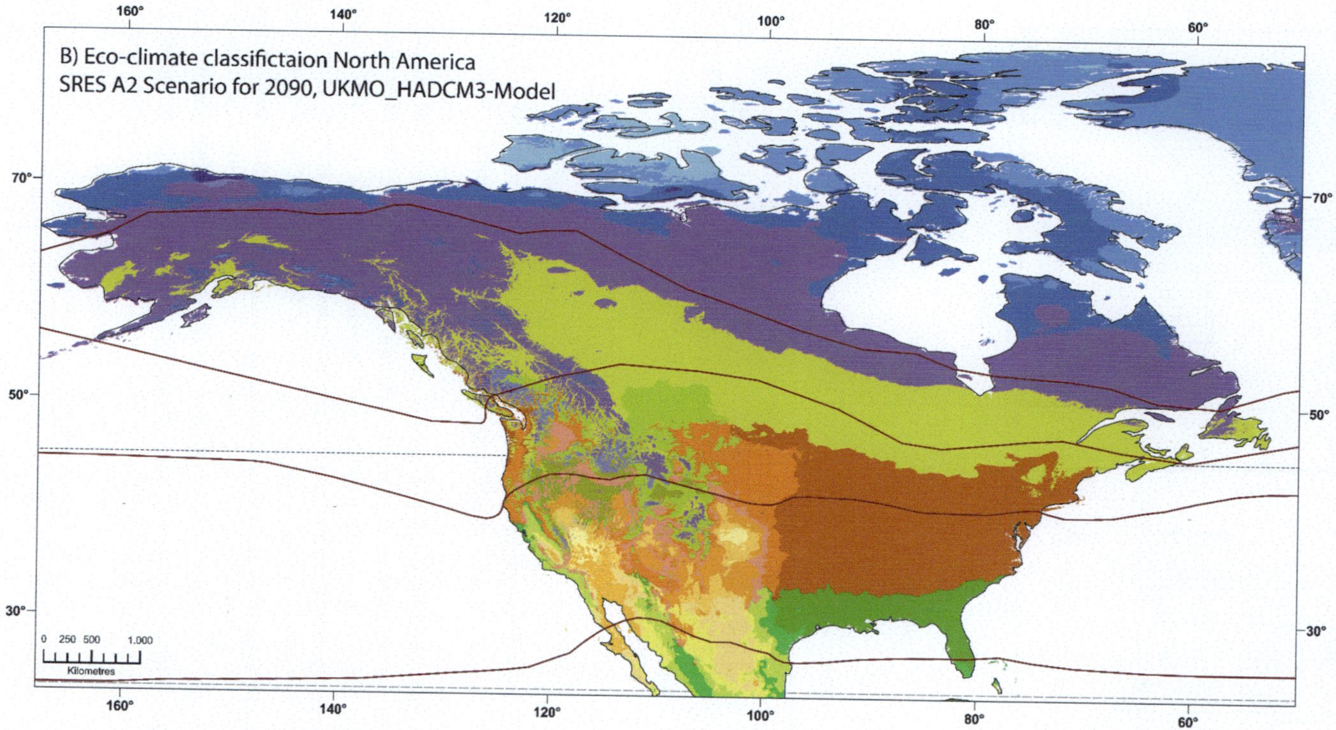

Fig. 11.1 **b** Shift of ecoclimatic zones in North America by **A2** scenario 2090, HadCM3 model (cf. Fig. 11.1a). (The lines indicate the boundaries of main recent climate zones and perfectly express the shift of climate zones compared to today) (for legend see Figs. 4.8 and 4.9)

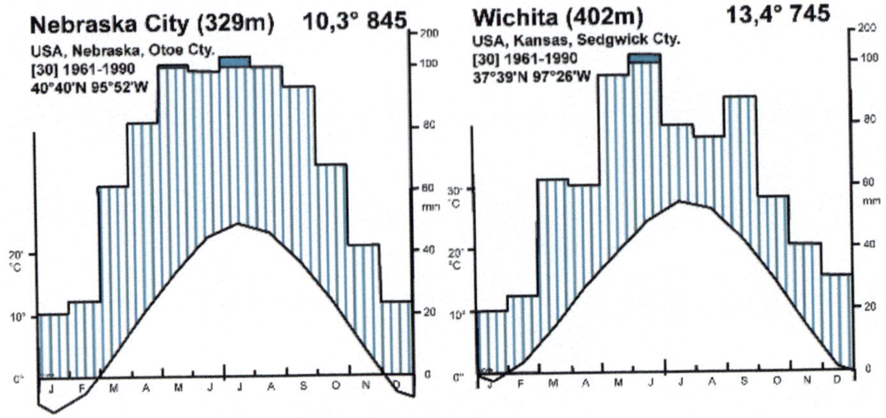

Fig. 11.2 "Twin-diagrams" of Nebraska City (Nebraska) and Wichita (Kansas) show how climate zones in western North America may shift due to climate change at the end of the 21th century

changes in the number of forest tree species. Over the next 100 years, the distribution areas of some taxa will shift around 500 km to the north. The sugar maple will probably disappear from the New England states (Melillo et al. 2002, 2011). In eastern North America, maple-beech-birch forests will gain in importance, while fir-spruce forests will largely retreat from this area (Iverson and Prasad 2001). The local climatic changes are expressed above all in increased winter temperatures. Precipitation generally falls as rain, and the areas with permanent winter snow cover are limited to smaller areas in the extreme north-east of this region (Frumhoff et al. 2007).

A simple comparison again can be made by "twin-diagrams" (Fig. 11.2). The climate of Nebraska City (Nebraska) will alter and become similar to the today's climate of Wichita (Kansas), which is about 360 km to the south. It will become warmer and probably less rain.

In the western United States, the effects of climate change at the end of the twenty-first century are particularly evident in mountainous regions. The oligotherm-humid climates of the nival belt are disappearing completely in favour of a microtherm-euhumid climate, which in turn is confined to narrow peaks and ridges.

Changes in precipitation that reduce the winter snowpack (Cayan et al. 2008) pose numerous problems for the arid western region of the continent. This is because both urban centres and the agricultural sector depend on the winter snowpack for their drinking and irrigation water supplies, respectively. The Sacramento Valley is the agricultural heartland between San Francisco and Los Angeles at the western slopes of the Sierra Nevada. Except for a very small area with a macrotherm-semiarid climate in the extreme south of the basin, the basin today has a macrotherm-subhumid subtropical climate. At the end of the twenty-first century, due to rising temperatures and decreasing precipitation, aridity will extend northwards from the southern part of the basin. This means that the northern part of the basin will also have a megatherm-semiarid climate. Towards the edges of the basin, however, the hygrothermal situation will become somewhat more favourable. In the Sierra Nevada, the microtherm or oligotherm climates of the high mountain belts will be replaced by climates of at least one higher thermal belt. The Great Basin area will come under megatherm-subhumid conditions. Further north in British Columbia, many conifer species are losing large portions of their natural range as a result of the disappearance of subboreal and montane climates (Hamann and Wang 2006). For example, the poleward shift of climate zones means that cedar-hemlock, douglas-fir, and ponderosa pine are colonizing the former sub-boreal and boreal ecosystems of the British Columbia high plateau (Hamann and Wang 2006).

11.2 Europe

With the exception of west-southwest France, the Po Valley, the Hungarian Basin and the Wallachia, a mesotherm-euhumid climate currently prevails throughout Europe with a latitudinal extension between the Alpine foothills and southern Sweden and a longitudinal extension from Ireland and central England to the Urals and south-western Siberia (Fig. 11.3a).

Taxon-based studies on the vegetation zone shifts in the course of climate change show the greatest changes for Europe between the Mediterranean areas and the Euro-Siberian region in the zonoecotones (Thuiller et al. 2005, 2006). Studies using dynamic vegetation models show that the greatest latitudinal and elevational landscape changes in Europe occur in the arctic and alpine ecosystems, where trees replace tundra vegetation or alpine mats respectively. Changes in the transition zone between temperate deciduous and boreal coniferous forest areas are particularly pronounced. In southern Europe, a change from forest to shrubland formations is expected due to increasing aridity (Hickler et al. 2012).

It is generally recognized that the interaction between land and atmosphere increases in Europe as greenhouse gas emissions increase (Koster 2004; Seneviratne et al. 2006), resulting in a northward migration of climate regions and the change in habitats for the flora (Asche and Stromberg 2024). These findings are consistent with the spatial arrangement of climate types reconstructed using the ecoclimate classification for the end of the twenty-first century.

Looking at the map for **A2 scenario** for Europe (Fig. 11.3b), central Europe will be covered by a macrotherm-euhumid climate of a thermophilic mixed deciduous forest formation between the edge of the Alps and the coasts of the North and Baltic Seas at the end of the twenty-first century. This type of climate almost completely will replace the mesotherm-euhumid climate with corresponding vegetation formation even in southern Ireland and southern England. In southern Finland, Jyväskylä (Fig. 11.4) with *Pinus*-taiga around will in some decades exhibit a climate like Vilnius today with a deciduous forest climate with *Quercus robur*, thus being twin-diagrams.

With the exception of the region around Brest, the entire territory of France will have a macrotherm-euhumid climate with sclerophyllous forest/shrubland of Mediterranean character. This is a consequence of the northward shift of the regions

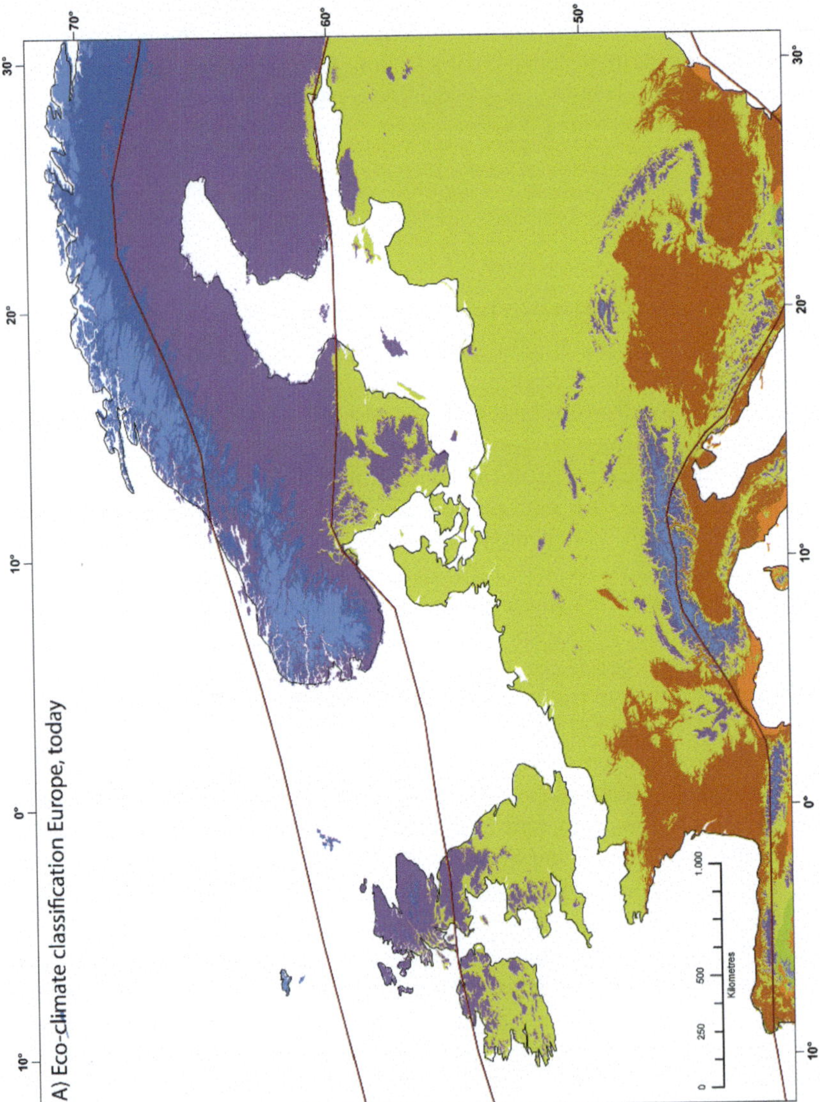

Fig. 11.3 **a** Ecoclimate zones in Europe today (lines indicate boundaries of recent climate zones) (legend see Figs. 4.8 and 4.9)

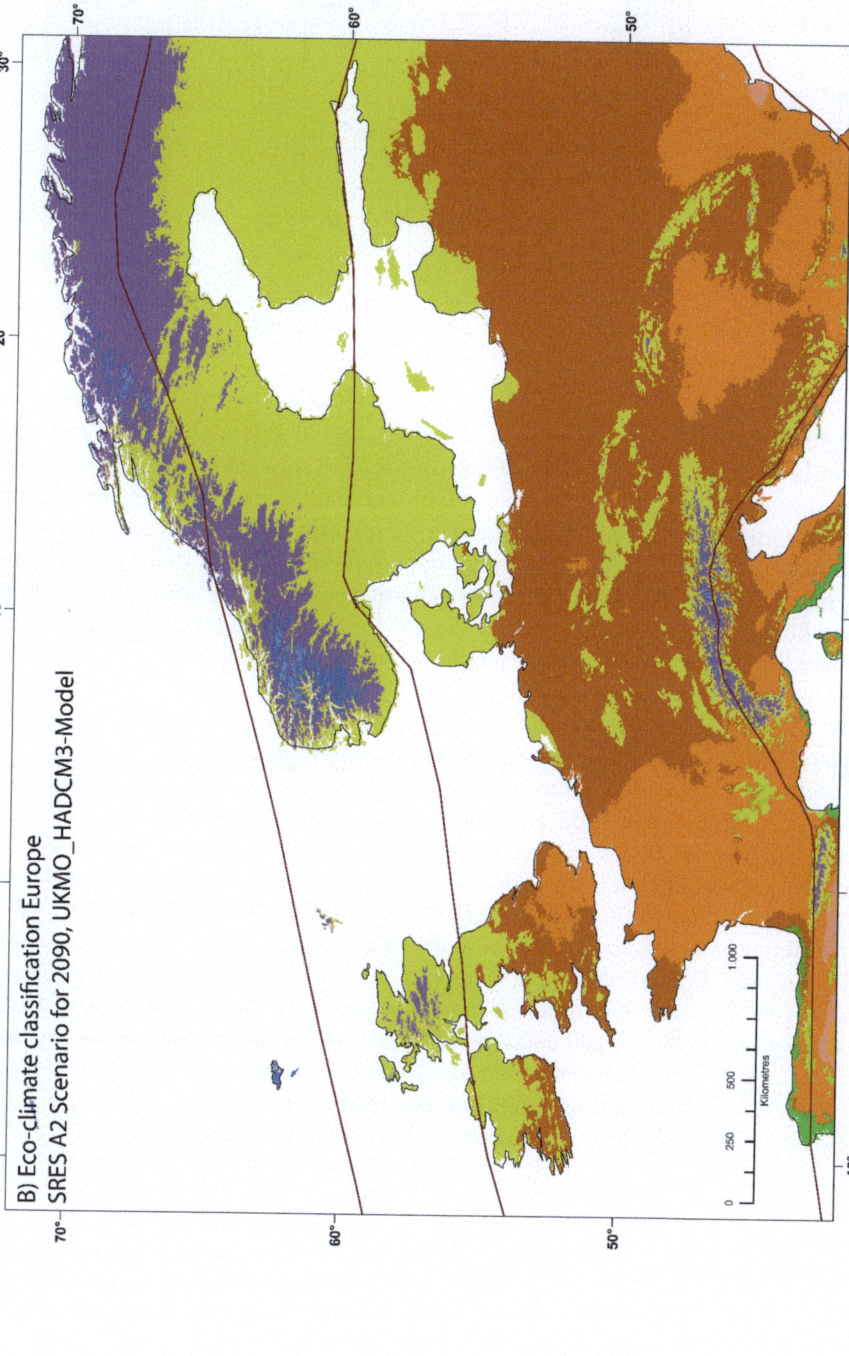

Fig. 11.3 **b** Shift of ecoclimate zones in Europe for A2 scenario 2090, HadCM3 model (cf. Fig. 11.3a) (Lines indicate boundaries of main recent climate zones) (for legend see Figs. 4.8 and 4.9)

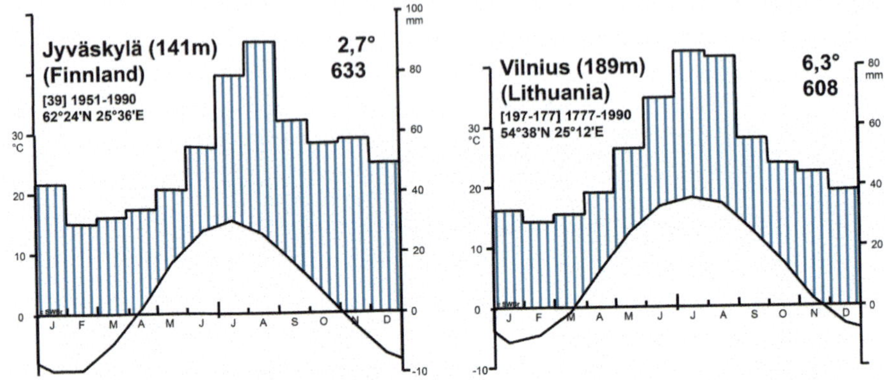

Fig. 11.4 "Twin-diagrams" of Jyväskylä (Finland) and Vilnius (Lithuania) show how climate zones in northern Europe may shift due to climate change at the end of the 21th century

of high soil-moisture-temperature coupling from the Mediterranean to central and eastern Europe (Seneviratne et al. 2006). In the high mountains of Europe (Pyrenees, Alps, Carpathians, etc.), the mesotherm-euhumid climate will also undergo a striking hypsometric change, replacing the microtherm-hyperhumid climate of the alpine mats. The areas of the hekistotherm-hyperhumid ice climate of the high mountains will be largely taken over by a microtherm-hyperhumid climate of the alpine mats, with the consequence of a drastic decline in the glaciated regions.

These findings of the ecoclimate classification are supported by microscale simulations of future climate change (Engler 2011), according to which 36 to 55% of alpine, 31 to 51% of subalpine and 19 to 46% of montane species in the mountains of central and southern Europe will lose around 80% of their habitats.

The drought in the Mediterranean region predicted for the end of the twenty-first century (Van Minnen et al. 2002; De Castro et al. 2007) may spread to northern France and even across the Channel region to southern England and southern Ireland due to a decrease in precipitation in the form of a macrotherm-euhumid climate. Mesoscale and microscale changes in climate types for central Europe result from the small-scale differences in temperature rise and spatial precipitation variability (De Castro et al. 2007). As a rule of thumb, in areas where there is a change in more than one climate type, the ecosystem is considered endangered for the taxa established there (De Castro et al. 2007). Such endangered areas, apart from northern Scandinavia, cover about 12% of the total land area of Europe (De Castro et al. 2007). These and many other individual findings from Europe clearly underline the shifts in climate zones due to climate change represented by the ecoclimate classification.

11.3 Mediterranean and the Middle East

The subtropical winter rainfall region of the Mediterranean (Fig. 11.5a, b) is considered as transition zone between the dry climate of the Sahara and the humid climate of central Europe (Flohn 1957; Koster 2004; Giorgi and Lionello 2008). Its climatic peculiarities are controlled by the interaction of the climatic processes of the tropics and the westerly wind drift. The transitional character of this zone thus results from the seasonal change in atmospheric circulation, which gives rise to areas of constant and alternating climates (Flohn 1957; Lauer 1991; Lionello et al. 2006, Breckle and Rafiqpoor 2022; Rafiqpoor 2024). The intensification of the subtropical anticyclones at the end of the twenty-first century is coupled with the stabilization of air masses and the northward shift of the Atlantic cyclogenesis (Giorgi and Lionello 2008). The consequence of this is the considerable decrease in precipitation in the Mediterranean region, with the exception of the Mediterranean Alps, where winter precipitation even increases slightly (Giorgi and Lionello 2008).

Overall, the robust signals generated by **GCMs** prove that the Mediterranean regions are likely to be among the most vulnerable areas to climate change (Giorgi and Lionello 2008). Due to the high geodiversity of major mountain blocks in this region, there are small-scale orography-induced precipitation changes that cannot be captured by global and regional models (Giorgi and Lionello 2008). However, the ecoclimate classification model (Fig. 11.5a, b) is also able to trace the small-scale climate changes induced by the relief. This is why the subtropics with the young mountains appear on the map as the most climatically differentiated regions on Earth. In the western Mediterranean region, extreme climatic changes are expected on the Iberian Peninsula (see Fig. 11.5a, b). Large parts of the Iberian Peninsula will experience longer dry and hot periods in summer, similar to those currently experienced in Morocco. For example, the provinces of Castilla-León and Castilla-La Mancha will be particularly affected by a macrotherm-semiarid climate. The southern provinces of Extremadura and Andalusia will also have much drier climates by the end of the twenty-first century compared to today (see Fig. 11.5a, b). The southern coastal regions of the Iberian Peninsula and North Africa around the Gulf of Cádiz will be affected by tropical climatic conditions as a result of the northward migration of the ITCZ (Evans et al. 2004; Evans 2009). A particularly striking phenomenon, however, is the slight climatic favouring of the Circum-Mediterranean coastal regions due to windward effects at the topographical elevations, especially in North Africa and the Levant, as an intensification of the land-sea wind phenomenon leading to increased convection and precipitation formation (Evans et al. 2004; Evans 2009). This effect is expressed in the occurrence of a megatherm-euhumid climate in almost the entire Circum-Mediterranean coastal regions, with the exception of the flatter coastal sections off Libya and Egypt. However, such small increases in precipitation cannot compensate for the problem of water deficit in the region, as they are accompanied by a rise in temperatures. In addition, winter precipitation even in the higher elevations of the mountains (Sierra Nevada in Andalusia, Atlas and Lebanon Mountains) falls mainly in liquid form. This means that it is no longer available as

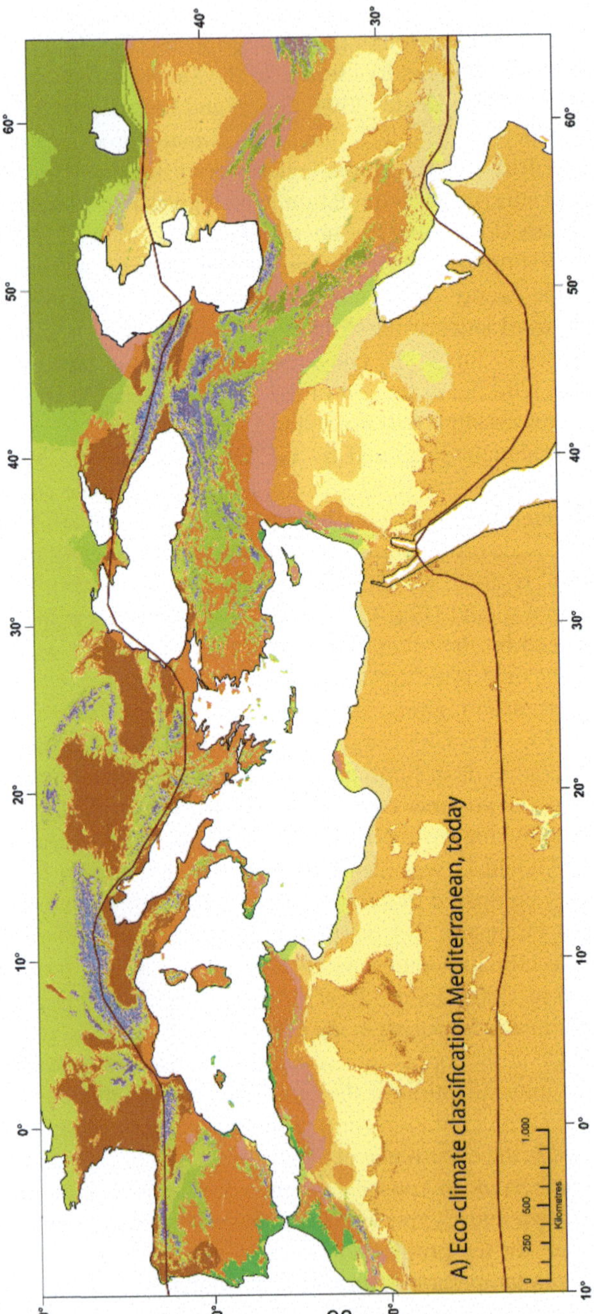

Fig. 11.5 **a** Ecoclimate zones in the Mediterranean (lines indicate the boundaries of the recent climate zones) (for legend see Figs. 4.8 and 4.9)

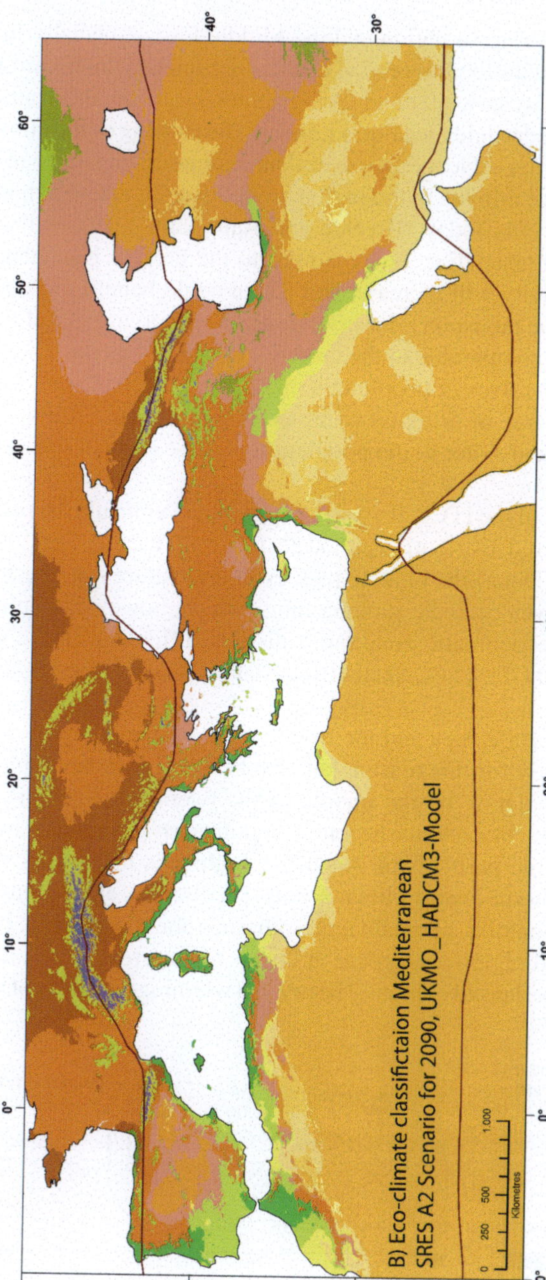

99

Fig. 11.5 b Shift of ecoclimatic zones in the Mediterranean (cf. Fig. 11.5a). **A2** scenario 2090, HadCM3 model (lines indicate boundaries of recent climate zones) (for legend see Figs. 4.8 and 4.9) a snow reserve for irrigation and drinking water supplies in the hot and dry summer months.

In the eastern Mediterranean region and the Middle East, especially in Iran and Afghanistan, which are at the extreme end of the Mediterranean winter rainfall region, where the cyclonic activity in winter and spring comes to an end, this region will be confronted with a severe drought (see Fig. 11.5a, b). The shifting of the climate zones in the Mediterranean region in Near East can be demonstrated using the twin diagrams of Kabul and Quetta. The climate in the region around Kabul, where a macrotherm-semihumid climate prevails today (Breckle and Rafiqpoor 2010), will change to a macrotherm-euarid climate (Fig. 11.6), similar to the present Kandahar climate. However, monsoon activity will increase, and Kabul could benefit more frequently from the slightly episodic monsoon rains in summer. In such a case, the climate of the Kabul region would be comparable to the current climate of Quetta. These changes will not only threaten the lives of people in a megacity in Afghanistan, where there is already a heavy shortage of drinking water today (Roostai and Rafiqpoor 2025), but the supply of food and water to the population of large parts of the country will become a major problem.

By the northern shift of the ITCZ also over the Arabian Peninsula the precipitation regime may have an altered distribution period despite similar annual amounts of rain. Thus, in the Emirates (Doha) the climate may exhibit in future decades less winter and more episodic or even periodic summer rains, which, however, would mean no visible change in the ecoclimatic zone but a distinct difference in the ecological climate situation (Fig. 11.7). Thus, Salala from South-Oman can be regarded as the twin-diagram.

The countries of North Africa and the Middle East will be severely affected by water shortages, not only for this reason, but also because of demographic growth, which will double population by the middle of the twenty-first century (Lionello et al. 2006). In Greece, Macedonia becomes significantly drier. The mesotherm-euhumid climate of large parts of the Anatolian highlands disappears completely in favour of a macrotherm-euarid climate; only small ridge and summit regions, especially in eastern Anatolia, maintain a mesotherm-euhumid climate. Even in the Caucasus, the glaciated regions disappear in favour of a narrow belt of alpine mats with a microtherm-hyperhumid climate. The dry and warming tendency of the climate

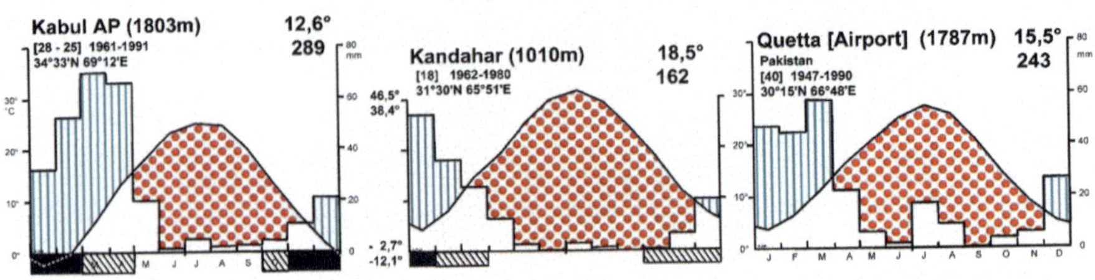

Fig. 11.6 Diagrams of Kabul, Kandahar and Quetta show how climate zones in Afghanistan may shift due to climate change at the end of the 21th century

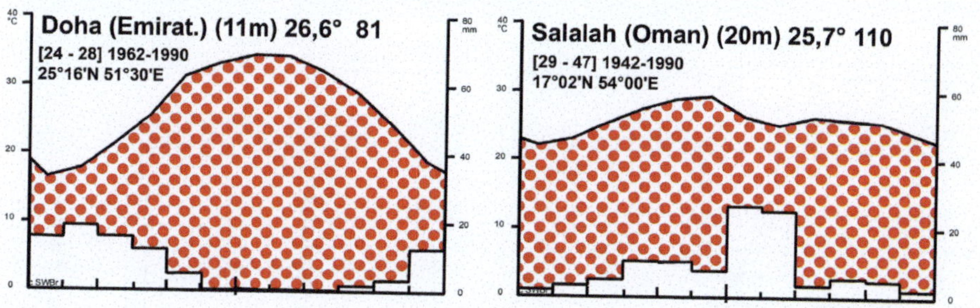

Fig. 11.7 Twin-diagrams of Salalah (Oman) and Doha (Emirate) show how climate zones in mediterranean deserts may shift due to climate change at the end of the 21th century

continues in the winter rainfall zone from the Mediterranean Sea via Iraq, Iran and Afghanistan to the Karakoram and western Himalayas.

11.4 Monsoon Regions in Asia (India, Japan, China)

In the monsoon regions of Asia, over 75% of annual rains occurs between June and September. They determine human life and the productivity of agricultural crops in this region. Since the two countries India and China alone are home to 1/3 of the world's population (Hamilton 1987; Kumar and Ashrit 2001; Taphyal and Rajeevan 2003), their alteration in the course of climate change is therefore crucial for human existence. In India, the annual variability and intensity of the summer monsoon are based on the interaction of various factors, e.g. the amount of precipitation, the vertical wind shear between pressure surfaces at different altitudes and the convection conditions in some regions (Dobler and Ahrens 2011; Rafiqpoor 2024).

The current picture of the ecoclimates in Asia (Fig. 11.8a) shows megathermal subhumid to semiarid conditions for large parts of the Indian subcontinent. In the southern Indian province of Goa and in Sri Lanka, but also in Assam and Bangladesh as well as on the slopes of the Himalayas as far as Kashmir, megatherm-euhumid and hyperhumid tropical climate types occur as a sign of pronounced and better distributed stratiform (monsoon) and cumuliform (convective) precipitation over the course of the year. The western lowlands of Rajasthan and Gujarat (Desert Thar) remain desert-like.

In the **A2** scenario (see Fig. 11.8a, b), the distribution areas of the circumscribed climate types in the lowlands of the Indian subcontinent remain almost unchanged compared to today. On the slopes of the Himalayas, however, the megatherm-euhumid tropical climate extends along the altitudinal gradient as a closed band from east to west. In particular, the warming of the Bay of Bengal, the longer residence time of humid air masses in the atmosphere and an intensification of convection (Dobler and Ahrens 2011) result in intensive precipitation in the megatherm-hyperhumid regions around Cherrapunji and Assam. The consequence of this is an expansion

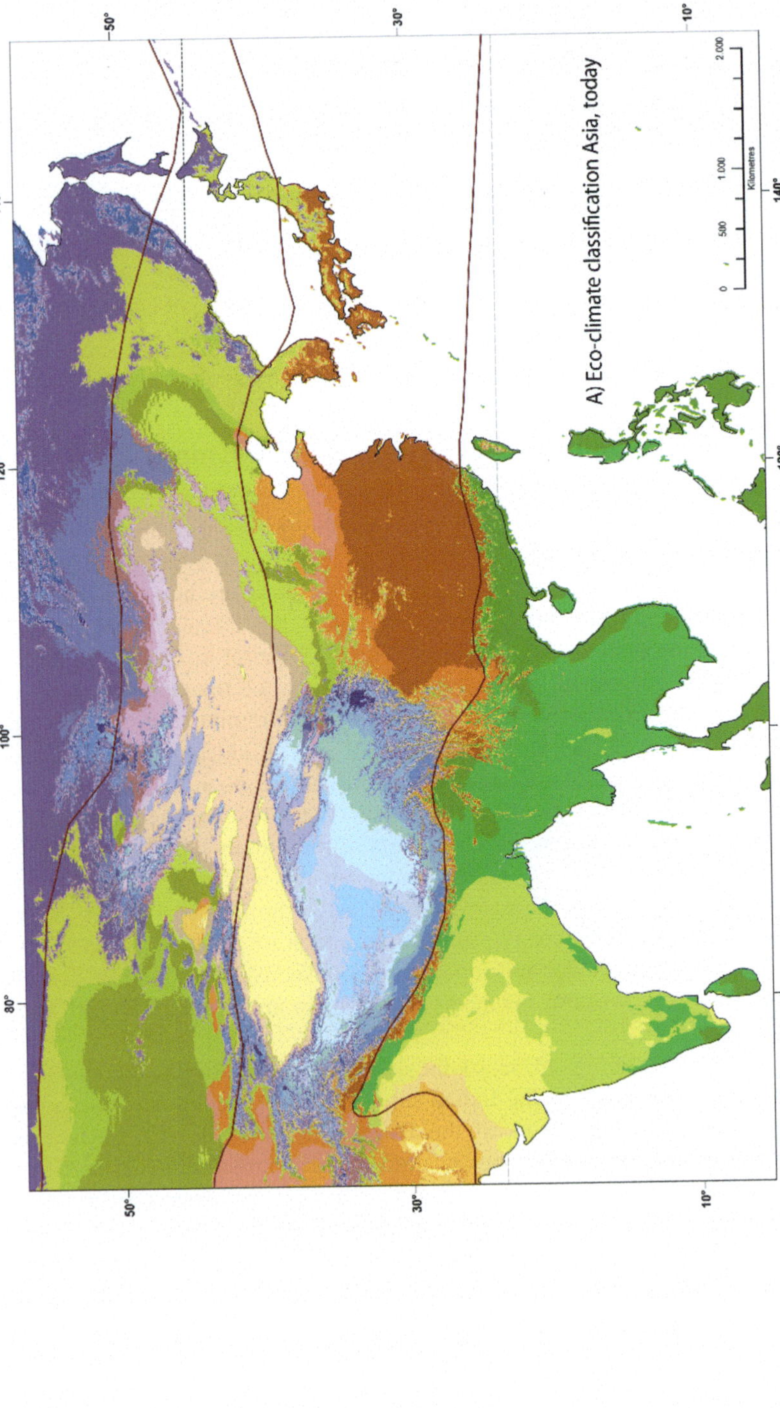

Fig. 11.8 **a** Ecoclimate zones in present-day Asia (lines indicate the boundaries of the recent climate zones) (for legend see Figs. 4.8 and 4.9)

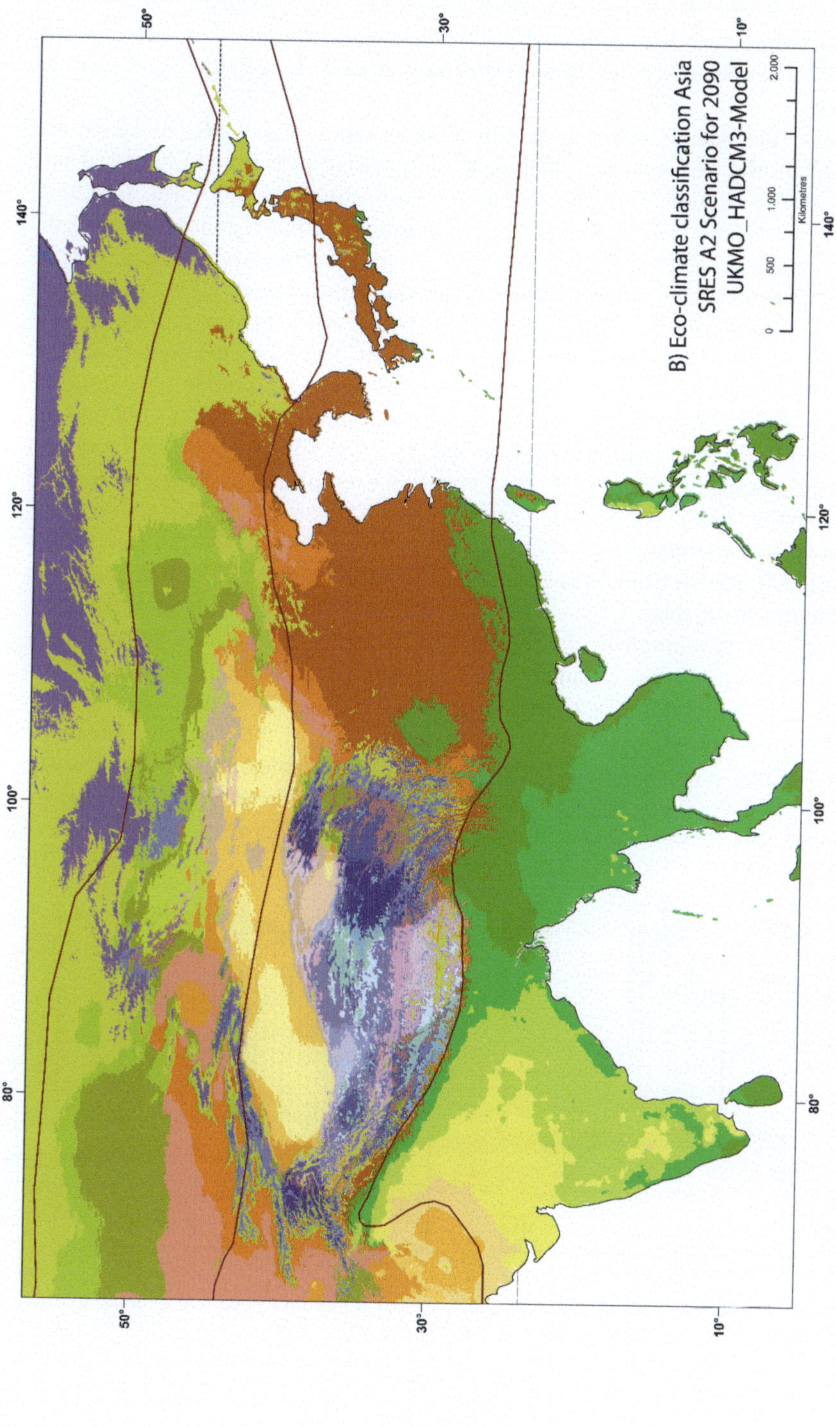

Fig. 11.8 b Shift of ecoclimatic zones in Asia (cf. Fig. 11.8a). A2 scenario 2090, HadCM3 model (lines indicate boundaries of recent climate zones) (for legend see Figs. 4.8 and 4.9)

of the areas of the megatherm-hyperhumid climate type in this region. The winter monsoon circulation also intensified at the end of the twenty-first century as a result of the rise in temperature and the associated increase in the thermal land-sea contrast. This affects the east coasts of India, especially around the Madras station in the south-east.

However, the spatial changes in the climate types under **SRES A2** of the **HADCM3 model** in China are particularly impressive (see Fig. 11.8a, b). There is a shift in the climate zones by around 5–6° of latitude to the north. In China, too, the intensification of the monsoon circulation is seen as a consequence of the increasing thermal land/water contrast, with the temperature rise at the end of the twenty-first century being greater in summer than in winter (Bueh 2003). In northern China, the microtherm-euhumid climate of the boreal zone is being replaced by the mesotherm-euhumid climate of the mixed deciduous forests of the cool-temperate latitudes. The latter gives way to a macrotherm-hyperhumid climate type from the subtropical region of the south. The winter temperatures become less cold and the monsoon summer rains will be stronger, the Yanzhou situation will be more and more replaced somehow by the climatic situation exemplified by the twin diagram of Yichang (Fig. 11.9).

The areas of the tropical megatherm-hyperhumid climate are also shifting further northwards in southern China. However, their spatial extent is very limited compared to the changes in northern China. This is a consequence of the only slightly intensified monsoon circulation in this region (Bueh 2003). The effects of the intensification of the monsoon circulation are not limited to northern China. Due to the increasing precipitation activity in the Sichuan Basin and the upper reaches of the Hunaghe River (Bueh 2003). The eastern slopes of the Tibetan Plateau are also affected by excessive small-scale, but overall favourable climatic changes (see Fig. 11.8a, b). Korea and Japan are almost completely affected by a macrotherm-hyperhumid subtropical climate. It even touches the southern tip of the island of

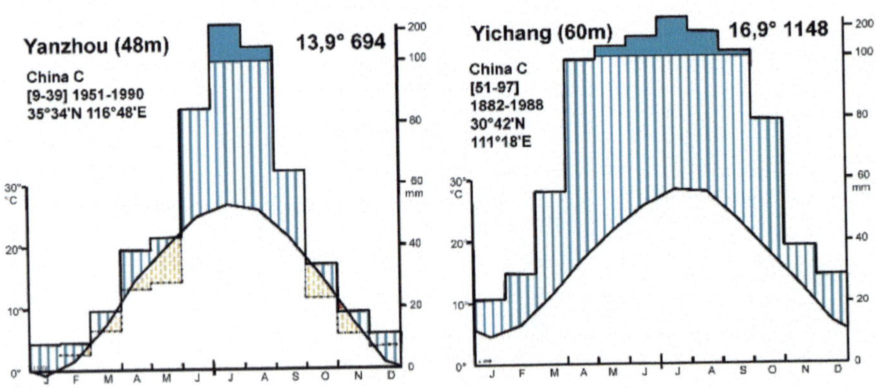

Fig. 11.9 Twin-diagrams of Yanzhou and Yichang (both China) show how climate zones in China may shift due to climate change at the end of the 21th century

Hokkaido, where today's microtherm-hyperhumid climate is replaced by mesotherm-hyperhumid climate conditions and the boreal coniferous forests are replaced by mixed deciduous forests.

11.5 The Arctic

Model simulations show that the Arctic and the high mountain regions of Europe, where tundra-like vegetation is being replaced by shrub and tree formations, represent the "hotspots" of landscape changes in the course of climate change (Hickler et al. 2012).

This was the result of studies with the "Generalized Dynamic Vegetation Model" based on IPCC SRES **A2** Emission scenario of the HadCM3 model (Hickler et al. 2012). Hickler et al. (2012) show that around 31–46% of Europe's total vegetation formations will change their appearance by 2085. If climate change continues to stabilize, as predicted by the IPCC SRES **A2** emissions scenario, vegetation formations would be expected to change by 76–80% by the end of the twenty-first century, as vegetation would then have sufficient time to gradually adapt to the changing climate conditions.

Past and present situation of the Arctic terrestrial and marine environment are given by Kassens et al. (1999) for the Siberian Arctic. This can be used as a baseline for future changes, since compared to all other climate zones, the most striking landscape changes occur in the high latitudes (boreal, subpolar and polar regions) of the northern hemisphere (Beaumont et al. 2011; Rogers et al. 2012) (Fig. 11.10a, b). In the circumpolar region of the northern hemisphere, large areas of oligotherm-hyperhumid tundra climate will give way to the microtherm-hyperhumid climate of the boreal coniferous forests at the end of the twenty-first century. The tundra climate zone will retreat to small areas north of 70° latitude, and the hekistotherm ice climate of the polar regions will disappear almost completely from the continental regions of the northern hemisphere. The Scandes will be dominated over a large area by the microtherm-hyperhumid climate of the boreal coniferous forests. The mesotherm-humid climate of the mixed deciduous forests of the cool mid-latitudes will extend beyond the 70th parallel in the fjords of Norway and even will reach the coasts of the Arctic Ocean in the lee of the Scandes around the bay of the White Sea and the Barents Sea. The oligotherm-hyperhumid tundra climate only will be restricted to tiny areas at limited altitudes in the Jotunheimen massif in Norway.

In boreal North America, similar changes will take place until the end of the twenty-first century. Here, too, the tundra will shrink to a narrow strip along the coasts of the Arctic Ocean, and the forests of the more southerly latitudes (mixed deciduous forests of the mid-latitudes and boreal coniferous forests) will advance latitudinally northwards by some 5–6 degrees of latitude. The hekistotherm polar ice climate will retreat completely from continental North America and is concentrated on Greenland and on the mountains of some islands in the Arctic Ocean and around Baffin Bay under the influence of the cold Labrador Current.

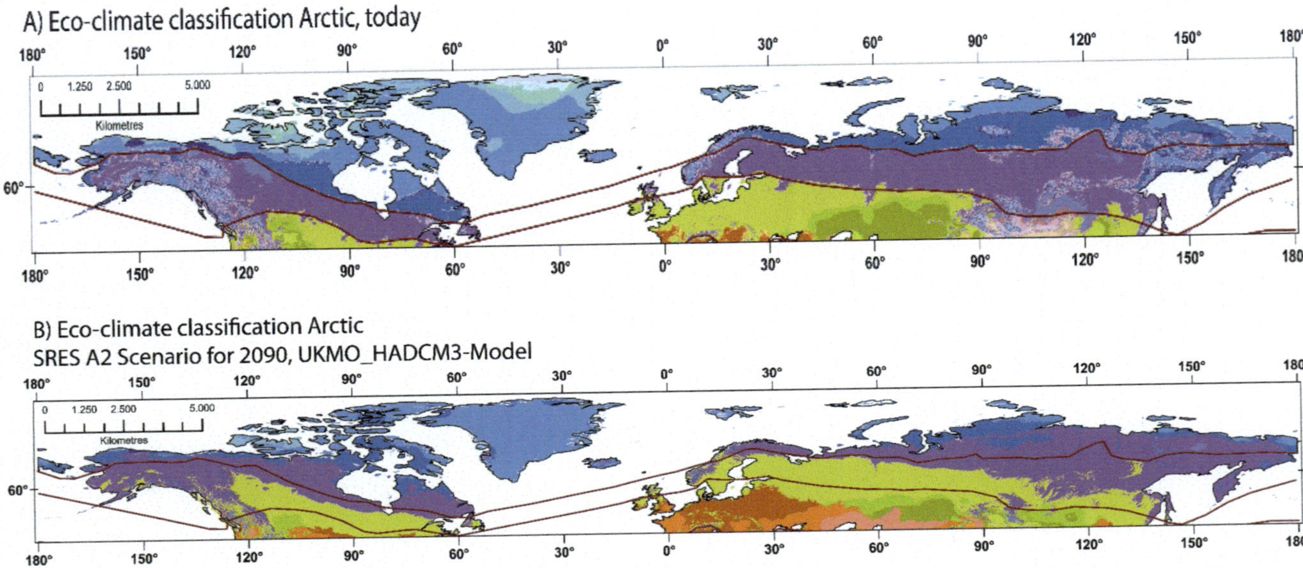

Fig. 11.10 **a** Ecoclimate zones in the Arctic today (lines indicate the boundaries of the recent climate zones) (for legend see Figs. 4.8 and 4.9). **b** Shift of ecoclimate zones in the Arctic (compared to today: see Fig. 11.10a) for A2 Scenario 2090, HadCM3 Model (lines indicate the boundaries of the recent climate zones) (for legend see Figs. 4.8 and 4.9)

11.6 The Tropics

The tropics (i.e., the tropical zone) are geographically located between the Tropics of Cancer and Capricorn with megatherm climates of different hygric conditions all year round. The differentiation of climate types in the tropics is therefore based on hygric criteria (Lauer 1975; Breckle and Rafiqpoor 2022; Rafiqpoor 2024). Depending on the length of the hygric growing season, in the tropics we can distinguish between a range of climate types from hyperhumid to hyperarid tropical climates in a three-dimensional arrangement (see Table 4.8 and 4.1; Fig. 4.8 and 4.11a, b; see also Fig. 4.8 and 4.10, ET-01 to ET-06).

A comparison of the map of today's ecoclimate types (Fig. 4.11a, b) with that of the **HADCM3 model** for the **SRES A2** scenario around 2090 (Fig. 4.11a, b) shows that there are hardly any climatic changes, especially in the equatorial regions of Africa and Asia. Only in South America the Amazon rainforest will be shrinking slightly in favour of a megatherm-euhumid savannah climate, and a corridor with megatherm-euhumid to subhumid savannah climates will open up between Santarém (Brazil) and Roraima (Venezuela). The area of megatherm-hyperhumid climate around the Tropic of Capricorn will extend as far south as the Bay of Buenos Aires (see Fig. 4.11a, b). The Llanos and the arid region of north-eastern Brazil will become even drier. In the area of the Central American land bridge, especially in Guatemala and on the Yucatán Peninsula, the drought season will become more excessive. In North America, a broad strip along the Gulf Coast north of the Tropic of Cancer will be occupied by a megatherm-hyperhumid tropical climate. However, strong climatic and landscape changes will occur worldwide in the tropical high mountains in the altitudinal structure, where the rise in temperature by the end of the twenty-first century is causing the glaciers to shrink as water towers in these regions. We can only imagine what problems will arise for megacities in the tropics in terms of drinking water supply as population pressure increases.

In Africa, hardly any changes in the spatial extent of the megatherm-hyperhumid equatorial rainforest climate may be observed in the Congo Basin. The obvious intensification of the "*predicted increase in wind stress over the southern Atlantic due to global climate change*" (Lutjeharms et al. 2001) inevitably will lead to an intensification of the Benguela Current on the west coast of Africa combined with an intensification of the upwelling phenomenon. The consequence of this will be an intensification of the cold upwelling water with excessive drought along the west coast, which may extend to the mouth of the Congo River. The Ethiopian highlands and the mountainous region around the Rift Valley in East Africa will become somewhat wetter due to the intensification of the monsoon circulation. On the east coast of southern Africa, the macrotherm-euhumid climate of the Maputoland-Pondoland diversity centre (Barthlott 2005) will be replaced by a tropical megatherm-euhumid climate. This is a result of the increased flow of warm water into the Agulhas Current as a consequence of the intensification of atmospheric circulation and wind patterns over the Indian Ocean water surface, leading to pronounced rainfall on the eastern coasts of South Africa (Lutjeharms et al. 2001). The coastal strip from Durban to

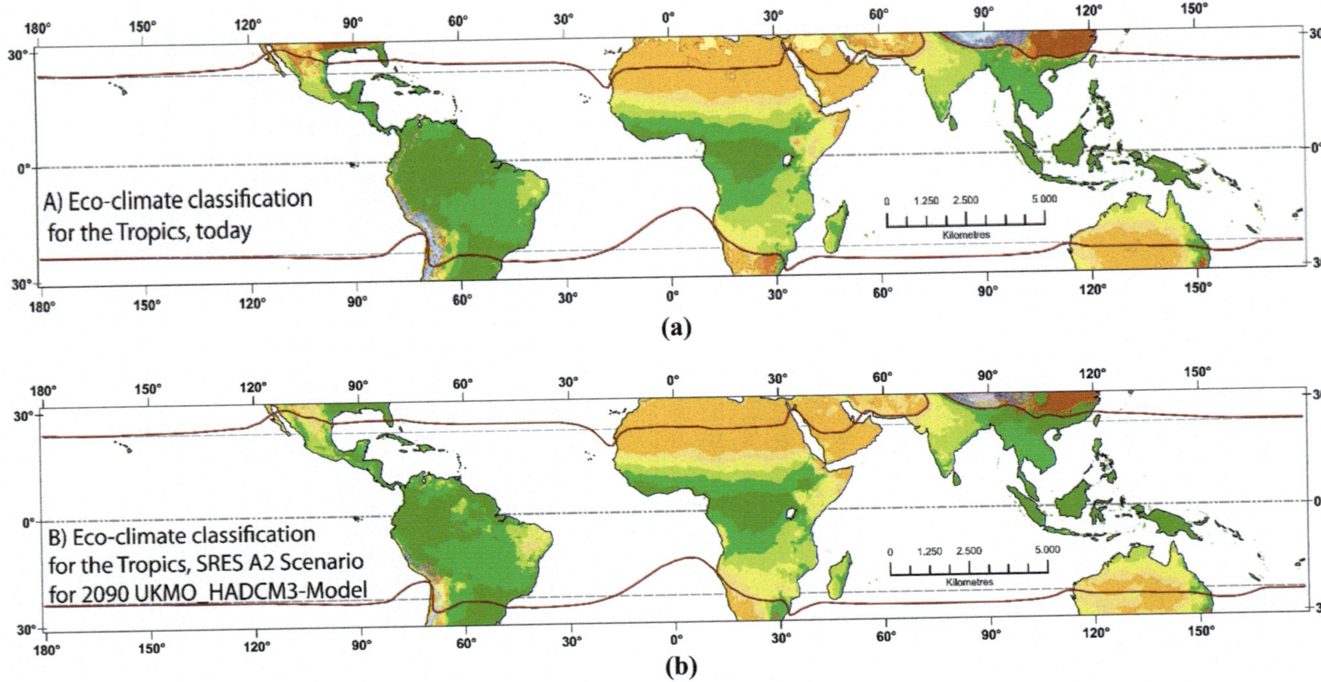

Fig. 11.11 **a** Ecoclimate zones in the tropics today (lines indicate the boundaries of the recent climate zones) (for legend see Figs. 4.8 and 4.9). **b** Shift of ecoclimate zones in the tropics (compared to today: see Fig. 11.11a) for A2 Scenario 2090, **HadCM3 Model** (lines indicate the boundaries of the recent climate zones) (for legend see Figs. 4.8 and 4.9)

Port Elizabeth therefore will take on a rainforest character, while the leeward inland of the entire high plateau in southern Africa will become drier.

References

Asche N, Stromberg H (2024) Wald verändert sich. Schafft der Klimawandel Raum für Baumarten aus anderen biogeografischen Regionen? In: Stadler, J (ed) Biodiversität und Klima—Vernetzung der Akteure in Deutschland XIX. Dokumentation der 19. Tagung. Bundesamt für Naturschutz. BfN–Schriften 694: 59–62

Barthlott W, Mutke J, Rafiqpoor MD, Kier G, Kreft H (2005) Global centres of vascular plant diversity. Nova Acta Leopoldina 92:61–83. https://www.researchgate.net/publication/215672852_Global_centers_of_vascular_plant_diversity

Beaumont LJ, Pitman A, Perkins S, Zimmermann NE, Yoccoze NG, Thuiller W (2011) Impacts of climate change on the world's most exceptional ecoregions. PNAS 108(6):2306–2311. https://doi.org/10.1073/pnas.1007217108

Breckle S-W, Rafiqpoor MD (2010) Field Guide Afghanistan – Flora and Vegetation. Scientia Bonnensis, Bonn – Manama – New York – Floríanapolis. Druckcenter Meckenheim (DCM). ISBN 978-3-940-766-30-4.863 pp

Breckle S-W, Rafiqpoor MD (2022) Vegetation and climate. Springer, Heidelberg, New York. ISBN 789-3-662-64035-7

Bueh C (2003) Simulation of the future change of east Asian monsoon climate using the IPCC series A2 and B2 scenarios. Chin Sci Bull 48(10):1024–1030. https://doi.org/10.1007/BF03184220

Cayan DR, Maurer EP, Dettinger MD, Tyree M, Hayhoe K (2008) Climate change scenarios for the California region. Climatic Change, vol 87. Suppl 1:21–42. https://doi.org/10.1007/s10584-007-9377-6

De Castro M, Gallardo C, Jylha K, Tuomenvirta H (2007) The use of a climate-type classification for assessing climate change effects in Europe from an ensemble of regional climate models. Clim Change 81:329–341. https://doi.org/10.1007/s10584-006-9224-1

Dobler A, Ahrens B (2011) Four climate change scenarios for the Indian summer monsoon by the regional climate model COSMO–CLM. J Geophys Res Atm 116(D24):24104: 1–13. https://doi.org/10.1029/2011JD016329

Engler REA (2011) 21st century climate change threatens mountain flora unequally across Europe. Glob Change Biol 17(7):2330–2341. https://doi.org/10.1111/j.1365-2486.2010.02393.x

Evans JP (2009) 21st century climate change in the Middle East. Clim Change 92:417–432. https://doi.org/10.1007/s10584-008-9438-5

Evans JP, Smith RB, Oglesby RJ (2004) Middle East climate simulation and dominant precipitation processes. Int J Climatol 24:1671–1694. https://doi.org/10.1002/joc.1084

Flohn H (1957) Zur Frage der Einteilung der Klimazonen. Erdkunde 11:161–175

Frumhoff PC, McCarthy JJ, Melillo JM, Moser SC, Wuebbles DJ (2007) Confronting climate change in the U.S. Northeast: science, impacts, and solutions. Synthesis report of the Northeast Climate Impacts Assessment (NECIA). Cambridge, MA: Union of Concerned Scientists (UCS). https://www.esf.edu/glrc/library/documents/ConfrontingClimateChangeintheUSNortheast_2007.pdf

Giorgi F, Lionello P (2008) Climate change projections for the mediterranean region. Global Planet Change 63:90–104. https://doi.org/10.1016/j.gloplacha.2007.09.005

Hamann A, Wang T (2006) Potential effects of climate change on ecosystem and tree species distribution in British Columbia. Ecology 87(11):2773–2786. https://doi.org/10.1890/0012-9658(2006)87[2773:PEOCCO]2.0.CO;2

Hamilton MG (1987) Monsoons—an introduction. Weather 42:186–193

Hickler Th, Bolte A, Hartrad B, Beierkhunlein C (2012) Folgen des Klimawandels für die Biodiversität in Wald und Forst. In: Mosbrugger V, Brasseur GP, Schaller M, Stribrny B (eds) Klimawandel und Biodiversität: Folgen für Deutschland. Chapter 8: 164–221. Wissenschaftliche Buchgesellschaft, Darmstadt. https://doi.org/10.13140/2.1.4759.9049

Iverson LR, Prasad, AM (1998) Predicting abundance of 80 tree species following climate change in the eastern United States Ecological Monographs 68:465–485

Iverson L, Prasad AM (2001) Potential changes in tree species richness and forest community types following climate change. Ecosystems 4(3):186–199. http://www.jstor.org/stable/3658952

Iverson LR, Prasad AM, Matthews S (2008) Modeling potential climate change impacts on the trees of the northeastern United States. Mitig Adapt Strateg Glob Chang 13:487–516. https://doi.org/10.1007/s11027-007-9129-y

Kassens H, Bauch HA, Dmitrenko IA, Eicken H, et al (eds) (1999) Land-ocean system in the Siberian Arctic, dynamics and history. Springer, Berlin

Koster RDEA (2004) Regions of strong coupling between soil moisture and precipitation. Science 305(5687): 1138–1140. https://doi.org/10.1126/science.1100217

Kumar KR, Ashrit RG (2001) Regional aspects of global climate change simulations: validation and assessment of climate response over Indian monsoon region to transient increase of greenhouse gases and sulphate aerosols. Mausam 52:229–244

Lauer W (1975) Vom Wesen der Tropen—klimaökologische Studien zum Inhalt und zur Abgrenzung eines irdischen Landschaftsgürtels. Abhandlungen der Math.-nat. Klasse, Akademie d. Wissenschaften und der Literatur, Mainz, Jg. 3: 52 pp. https://www.adwmainz.de/wilhelm-lauer-stiftung/publikationen.html

Lauer W (1991) Climatology. In: Pancel L (ed) Tropical forestry handbook. Springer Verlag, Berlin, Heidelberg, New York, pp 95–164

Lionello PEA, Malanotte-Rizzoli P, Boscolo R (2006) Mediterranean climate variability. Elsevier. https://www.researchgate.net/publication/200472204

Lutjeharms JRE, Monteiro PMS, Tyson PD, Obura D (2001) Oceans around Southern Africa and regional effects of global change. South African J Sci 97: 119 ff. https://www.researchgate.net/publication/30509213

Melillo JM, Steudler PA, Aber JD, Newkirk K, Lux H et al (2002) Soil warming and carbon-cycle feedbacks to the climate system. Science 298(5601):2173–2176. https://doi.org/10.1126/science.1074153

Melillo JM, Butler S, Johnson J, Mohan J, Steudler P et al (2011) Soil warming, carbon–nitrogen interactions, and forest carbon budgets. PNAS 108(23):9508–9512. https://doi.org/10.1073/pnas.1018189108

Rafiqpoor MD (2024): Climatology. اقلیم شناسی—DCM-Druck Center Meckenheim. Germany, pp 298. https://www.researchgate.net/publication/377066161_Climatology_-_aqlym_shnasy

Rogers BM, Randerson JT, Bonan GB (2012) High-latitude cooling associated with landscape changes from North American boreal forest fires. Biogeosciences Discussions 9(9):12087–12136. https://doi.org/10.5194/bgd-9-12087-2012

Roostai AH, Rafiqpoor MD (2025) Problems of water scarcity in the city of Kabul – The importance of the Kârez's regeneration in Afghanistan. DCM-Druck Center Meckenheim. ISBN 978-3000-803-956. https://www.researchgate.net/publication/388146532_Problems_of_water_scarcity_in_the_city_of_Kabul_-_The_importance_of_the_Karez's_regener-ation_in_Afghanistan

Seneviratne SI, Lüthi D, Litschi M, Schär C (2006) Land-atmosphere coupling and climate change in Europe. Nature 443:205–209

Shriner DS, Street RB (1998) North America. In: Watson RT, Zinyowera MC, Moss RH (eds) The regional impacts of climate change. Cambridge Univ. Press, New York, pp 253–330

Taphyal V, Rajeevan M (2003) Updated operational models for long-range forecasts of Indian summer monsoon rainfall. Mausam 54:495–504

Thuiller W, Lavorel S, Araújo MB, Sykes MT, Prentice IC (2005) Climate change threats to plant diversity in Europe. PNAS 102:8245–8250

Thuiller W, Lavorel S, Sykes MT, Araujo MB (2006) Using niche-based modelling to assess the impact of climate change on tree functional diversity in Europe. Diversity Distrib 12:49–60

Van Minnen JG, Onigkeit J, Alcamo J (2002) Critical climate change as an approach to assess climate change impacts in Europe: development and application. Environ Sci Policy 5:335–347. https://doi.org/10.1016/S1462-9011(02)00044-8

Conclusions

The ecoclimate classification proves to be a useful tool for reconstructing the global shifts in climate zones due to climate change. The predictions of climatic changes reconstructed by the 'Climate Envelope' models are far exceeded by the synthetic character of the ecoclimate classification due to the use of ecologically relevant parameters. The ecoclimate classification shows a significant shift of the climate zones on the northern hemisphere up to 5–6° of latitude polewards by the end of the twenty-first century. Major changes will occur in the high and mid-latitudes of the northern hemisphere. They are most noticeable at the zonoecotones and in the orobiomes, while the central parts of the zonobiomes will remain almost climatically constant. In the polar regions, the icy regions may disappear from the continental areas of the northern hemisphere, with the exception of central Greenland, the edges of which will also become ice-free over large areas. The tundra regions will be almost completely taken over by shrub and tree vegetation, and the boreal coniferous forests will make the way free for the mixed deciduous forests of the mid-latitudes in their southern margins. At the end of the twenty-first century, central Europe may be dominated by a macrotherm-hyperhumid climate with a thermophilic deciduous forest formation. In the Mediterranean regions, the drought will intensify, especially on the Iberian Peninsula. The peripheral mountains of the Circum-Mediterranean are favoured hygrically by windward effects, but winter precipitation may fall as rain at the expense of snow accumulation with major consequences for the supply of drinking and irrigation water in hot summer months. In Asia (India, China, Japan, Korea), the intensification of the monsoon circulation will lead to pronounced shifts in climate zones. Above all, the dry areas of northern China will benefit from the future climate, despite mankind sometimes being put under pressure by extreme events such as heavy rainfall, flooding and typhoons. In the tropics, the high mountains will experience a pronounced altitudinal change in climate and vegetation, while the effects of climate change will barely be noticeable in the lowlands of the humid tropics. The use of PET data from the present climatology to reconstruct the climates of the future under climate change undoubtedly underestimates the real extent of climate

M. D. Rafiqpoor and S.-W. Breckle, *The World's Ecoclimates*,
SpringerBriefs in Earth System Sciences,
https://doi.org/10.1007/978-3-032-06456-1

zone shifts. Therefore, there is an urgent need for research to develop a **database for "PET for future climatology"**, as is the case for temperature and precipitation (https://www.worldclim.org/data/cmip6/cmip6climate.html), to calculate more accurately also the water balance for modelling future climates. Nevertheless, our ecoclimate maps for the selected time slices and **GCM** models show a convincing picture of the shifts in climate zones, which are congruent with the findings of various studies from the individual regions of the earth.

Summary

The ecoclimate classification and associated dataset is a promising approach for exploration of broad-scale patterns in the distribution of species, plant functional types, zonobiomes, as well as patterns in species richness. It can also be used to place local or regional studies in a global ecoclimatic context. In addition, the data can be used to study projected changes in ecoclimate and vegetation due to global climate change. The classification takes into account the potential natural vegetation as a reference base via the heat and water balances. This makes it easy to reconstruct the changes in vegetation caused by global climate change. It can also be used to reconstruct changes in global climate caused by changes in vegetation cover. The model can also be used to calculate and map the changes in vegetation cover under climate change. Contributions from Working Group I to the Third Assessment Report of the Intergovernmental Panel on Climate Change predict continued atmospheric warming, with mean annual temperature increases of 1.8–3.5 K by 2050 and 2–7 K by 2100 (Cubasch 2001). Such changes are likely to result in spatially altered combinations of climatic and topo-edaphic factors that will have significant impacts on the distribution and persistence of natural vegetation and animal species (see Cubasch 2001; Hobbs et al. 2006; Ohlemüller et al. 2006; Ricciardi 2007; Williams and Jackson 2007; Williams et al. 2007). It is precisely in this context that the ecoclimate classification is a powerful measure to assess the dimensions of habitat change for flora and fauna, or conversely to infer from changes in vegetation cover the shifting of ecoclimatic boundaries.

References

Cubasch U (2001) Climate change 2001: the scientific basis. Third Assessment Report of IPCC, Chapter 9: 525–582. https://www.researchgate.net/publication/290809956.

M. D. Rafiqpoor and S.-W. Breckle, *The World's Ecoclimates*,
SpringerBriefs in Earth System Sciences,
https://doi.org/10.1007/978-3-032-06456-1

Hobbs RJ, Arico S, Aronson J, Baron JS, Bridgewater P et al (2006) Novel ecosystems: theoretical and management aspects of the new ecological world order. Glob Ecol Biogeogr 15(1):1–7. https://doi.org/10.1111/j.1466-822X.2006.00212.x.

Ohlemüller R, Gritti ES, Sykes MT, Thomas CD (2006) Towards European climate risk surfaces: the extent and distribution of analogous and non-analogous climates 1931–2100. Glob Ecol Biogeogr 15:395–405.

Ricciardi A (2007) Are modern biological invasions an unprecedented form of global change? Conserv Biol 21:329–336.

Williams JW, Jackson ST (2007) Novel climates, no-analogue communities, and ecological surprises. Front Ecol Environ 5(9):475–482. https://esajournals.onlinelibrary.wiley.com/doi/10.1890/070037

Williams JW, Jackson ST, Kutzbach JE (2007) Projected distributions of novel and disappearing climates by 2100 AD. PNAS 104(14):5738–5742. https://doi.org/10.1073/pnas.0606292104.

Index

M. D. Rafiqpoor and S.-W. Breckle, *The World's Ecoclimates*,
SpringerBriefs in Earth System Sciences,
https://doi.org/10.1007/978-3-032-06456-1

The manufacturer's authorised representative in the EU is Springer Nature Customer Service Centre GmbH, Europaplatz 3, 69115 Heidelberg, Germany. If you have any concerns regarding our products, please contact ProductSafety@springernature.com

Printed and bound by CPI Group (UK) Ltd, Croydon, CR0 4YY
21/09/2026
02219996-0002